And Others, W. H. P. Faunce

Searchlights on Christian Science

And Others, W. H. P. Faunce

Searchlights on Christian Science

ISBN/EAN: 9783337252595

Printed in Europe, USA, Canada, Australia, Japan

Cover: Foto ©berggeist007 / pixelio.de

More available books at **www.hansebooks.com**

Searchlights on Christian Science

A SYMPOSIUM

Fleming H. Revell Company

Chicago : New York : Toronto

1899

PREFACE

This series of articles upon Christian Science first appeared in THE STANDARD, of Chicago, during the first part of the present year. The original series, containing seven of the chapters, was prepared at the request of the editors to meet an evident demand. The choice of writers and assignment of topics was purposely so made as to insure a variety of method and of opinion. Had it been desired to produce a systematic treatise possessing entire unity and consistency, the task would naturally have been committed to a single writer. It was thought that the views of various men, some trained as students, others as practical pastors ministering to many classes, would better serve the purpose. This purpose was quite as much to inform THE STANDARD'S readers on the tenets and usages of Christian Science as to combat the new teaching. A certain divergence of feeling and of opinion due to the different estimates of the several writers is therefore not regarded as unfortunate. It fairly represents the sentiment of evangelical denominations toward Christian Science.

The last three articles were contributed to THE STANDARD as in some degree supplementary to

Preface

the others, emphasizing phases of the subject not fully treated before. Their "searchlights" are directed from new angles, and illuminate some untouched points. Taken as a whole, the series has already proved widely useful in correcting vague impressions and enabling the ordinary reader, unable or unwilling to peruse "Science and Health" and the other literature of Christian Science, to perceive the superficial resemblances and fundamental contrasts between Mrs. Eddy's doctrines and the teaching of Scripture. That it may, as now published, still further direct the thought of the Christian public to the true sources of knowledge concerning the duty and destiny of man, is the earnest hope of the authors.

August, 1899.

CONTENTS

Searchlights

ON

Christian Science

I

THE HISTORY OF CHRISTIAN SCIENCE

BY JOHN ROTHWELL SLATER

A religious movement which within a quarter
of a century has won many thousands of adher-
ents, without the slightest attractions in the way
of elaborate ritual or familiar creed, is worthy of
careful study. When that movement has drawn
many of its followers from evangelical churches,
including persons of unquestioned intelligence
and piety, such study becomes a duty. If the
new teaching be proved to contain elements of
truth which Christian people have ignored or
neglected, then we must correct our practice to
square with the truth, whether the truth be new
or old. If the novel doctrines prove to contain
fundamental errors, errors which are capable of
undermining Christian faith and of leading men
and women far from the simple gospel of the
New Testament, then those errors should be

9

plainly pointed out. This is not persecution; it
is not bigotry; it is not intolerance. Those words
all imply that the critics of the new doctrine
desire to restrict the social or civil liberties of
the persons whose views they seek to refute.
Baptists should be the last of all people to deny
to any man the widest liberty of conscience and
belief. Unless we are to proclaim ourselves
narrower than our forefathers, we shall continue
to demand for others the same civil rights and
privileges that we enjoy. We shall freely admit
the right of the Christian Scientist to those rights
and to our respectful and courteous attention.
Where his faith touches the law of the state, we
shall feel at liberty to demand his obedience to
that law, provided we believe the law to be a
salutary one for the protection of society. But
so far as his right to hold certain doctrines, and
to worship after his own preference, is concerned,
no thoughtful citizen can entertain a doubt, and
no true Christian can find warrant for giving way
to anger or abuse. The Christian Scientist is
justified in complaining that the very terms in
which some critics denounce his belief belie their
profession as followers of Jesus Christ, who
never denounced anybody except hypocrites.

Having said so much, the writer of this article,
and the writers of those which are to follow, will
not offer apology for any criticism, however
severe, passed upon the teaching and the practice
of Christian Science, considered as a system of
religion, philosophy and therapeutics. In that

aspect it is a legitimate subject of criticism; and if the objections offered are based upon correct statement of facts and correct principles of logic, they cannot be justly regarded as the outgrowth of bigotry.

THE GROWTH OF A FEW YEARS.

Christian Science has had a marvelous growth. Its origin was in the year 1866, when Mrs. Mary Baker G. Eddy first formulated the main doctrines which later developed into a system; but though she began teaching these doctrines in the following year, it was not until the publication of her text-book, "Science and Health," in 1875, that any considerable number of followers was attracted to the new faith. The first "Christian Scientist Association" was formed in Boston in 1876; this body three years later was incorporated as a church, entitled the First Church of Christ, Scientist. At its organization it had but twenty-six members; it has now about 1,400 resident members, besides more than 10,000 non-resident. The total number of Christian Scientists in the United States cannot be stated exactly. There are 304 chartered churches, a list of which may be found in the Christian Science Journal, a monthly magazine published by the parent church in Boston. In addition to these there are 111 regular Sunday services held by persons in sympathy with the denomination, many of which bodies will probably be organized as churches. The enrolled membership of the chartered churches is stated at 70,000. Leaders of the denomination claim, however, that the actual number of persons who regularly attend their services and accept their tenets is far in excess of this figure. The estimate of Mr. Carol Norton, one of the board of lecturers, who writes to the Independent of January 5, 1899, is "not less than 300,000 in the United States and Canada." While this figure will be regarded as excessive by many, there can be no doubt that Christian Science is growing rapidly. Until recently the Boston church was the only one with a building of its own, especially constructed for it, though congregations in many cities occupied rented quarters in the edifices of other denominations. Now, however, according to Mr. Norton, thirty buildings are in course of erection. The First Church of Christ, Scientist, in Chicago, dedicated November 14, 1897, free of debt, a beautiful marble structure, of Grecian architecture, on one of the finest boulevards of the South Side, costing over $100,000; and the Second and Third Churches, on the North and West Sides respectively, are planning to erect commodious and elegant buildings in the near future. The Second Church, of New York City, has begun the erection of a house of worship near Central Park, which will cost, according to the building permit, $220,000. Other building enterprises might be named.

The movement has grown most rapidly in the large cities, where it has attracted many persons of considerable means and social standing. Yet it has also pushed its way into scores and hundreds of towns and villages, where the numbers may be small and the meeting-place obscure, but the influence on evangelical churches considerable. Take, for instance, such a state as Iowa. In Iowa Christian Science churches or congregations are reported in no less than thirty-four cities and towns, as follows: Algona, Bunch, Burlington, Carpenter, Cedar Rapids, Charles City, Clinton, Council Bluffs, Creston, Davenport, Denison, Des Moines, Douds, Dubuque, Estherville, Exira, Fort Dodge, Independence, Keokuk, Le Mars, Marshalltown, Mason City, McGregor, Missouri Valley, Oskaloosa, Ottumwa, Rock Valley, Rudd, Sheldon, Sioux City, Tipton, Washington, Webster City, and What Cheer. In fourteen other towns of the state there are Christian Science practitioners, who of course disseminate more or less the doctrines by which they profess to cure disease. It is evident that, if this state is a fair example, the sympathizers with Christian Science in the entire country must be numerous, though perhaps 200,000 would be nearer the truth than 300,000.

And this has been the growth of twenty years. It does not fall within the province of this article to discuss the reasons for this remarkable increase, nor to decide whether it constitutes any serious menace to orthodox Christianity, nor to predict the future. A brief sketch of the first years of the movement, especially of Mrs. Eddy and her relation to it, will prepare us to consider in future articles questions of more vital interest to Christian pastors and people. For some of the facts here stated, the writer is indebted to an article by S. J. Hanna in Progress, a monthly magazine published by the University Association, Chicago, in June, 1898, containing a brief but clear account of the history and principles of Christian Science by an authorized exponent. Other facts have been derived from "Science and Health, with Key to the Scriptures," by Mrs. Mary Baker G. Eddy, the text-book of Christian Science, first published in 1875, now in its one hundred and sixtieth edition (of 1,000 copies each). This book is chiefly an exposition of the author's metaphysical system, with some practical suggestions on healing, the social relations, the interpretation of the scriptures, etc., but contains some historical references. Other information has been gained from the Christian Science Journal and various pamphlets published by the Christian Science Publishing Society, of Boston.

MRS. EDDY'S LIFE.

Mrs. Eddy's maiden name was Mary Baker. She was born near Concord, N. H., her parents being persons of Scotch and English extraction. The date of her birth is not stated, but as she was first married in 1843 it cannot be far from 1825. She received an academic education, followed by private study under the direction of tutors, including such subjects as natural philosophy, rhetoric, logic, metaphysics and the classics. While still very young she united with a Congregational church, and remained a member until 1879. She wrote for various magazines on literary and patriotic subjects and for many years took more or less part in certain philanthropic enterprises. Mrs. Eddy was twice married; her first husband was Col. G. W. Glover, her second Dr. A. G. Eddy. She studied homeopathy, but did not practice as a regular physician. During many years, according to her statement, she attempted to discover the true relation between physical ailments and mental conditions. In 1866, while suffering from an injury supposed to be incurable—the exact nature of which is not described—she applied the principle which she had finally formulated as to the unreality of disease and the power of "immortal mind" over the "error of mortal mind" commonly supposed to be disease; and the immediate cure to which she testifies led her to elaborate her theories, to undertake a study of the scriptures from her new point of view, and finally to write "Science and Health." For the instruction of others in the "principle and rule of spiritual science and metaphysical healing—in a word, Christian Science," she relied at first upon small classes. These grew into the Massachusetts Metaphysical College, incorporated in Boston in 1881, which still trains practitioners in the methods prescribed by the founder. The fees for instruction were large, the revenues from the sale of "Science and Health" also grew steadily,

and the movement began to assume large proportions.

Mrs. Eddy acted as pastor of the Boston church
from 1881 until 1895. She then retired, and now
lives on a fine country place near Concord, N.H.,
retaining a certain amount of oversight in the
direction of the affairs of the churches through
her communications to the Boston church and her
articles in the denominational magazine. While
no definite limits seem to be assigned to her
authority in this respect, her followers almost
universally regard her wishes as law, and have
no desire to dispute her commands. She is commonly referred to among them as "mother"; but
there seems to be no ground for the charge sometimes made by critics of the sect that she either
claims or allows her disciples to attribute to her
anything even approaching divine honors. The
charge may have arisen from the emphasis which
is laid in "Science and Health" upon what she
calls the divine maternity, the feminine element
in the divine nature.

THE PROMINENCE OF WOMEN IN CHRISTIAN SCIENCE.

In this connection a most remarkable feature of
the movement should be noted. It was not only
founded by a woman, but it has been spread
chiefly through the agency of women. The
practitioners, graduated from the training school
in Boston, who go all over the country to engage
in their occupation as healers and also to propa-
gate Christian Science doctrines, are nearly all

women.   In the Christian Science Journal (for
December, 1898) appears a list of registered practitioners numbering about 1,625; and of these
only about 290 are men, and nearly all of these
are the husbands of women whose names also
appear in the list. Most of the testimonies concerning cures that appear in the Journal (though
not all) are from women. While considerable
numbers of men attend the church services and
sympathize with the movement, it may be said,
so far as observation goes, that they are usually
attracted and interested through their wives and
families, rather than drawn to the churches of
their own accord by the character of the doctrines
there proclaimed.   Further, the men seem to lay
greater emphasis on the cures which they believe
to have been wrought by this agency than on the
abstract and often incomprehensible theories of
Mrs. Eddy's book.

CHURCH POLITY.

A brief account of the organization and form of
service of the Christian Science churches may
properly be included in this historical sketch.
These churches are incorporated like ordinary
churches, under state laws. They owe a certain
degree of loyalty to the parent church in Boston,
and contribute to the expenses of the movement,
but are for the most part self-governing. The
requirements for membership seem to be an
acceptance of the platform of the denomination,
containing a series of affirmations and negations

concerning mind, matter, truth, error, God and man, and a willingness to coöperate in spreading these doctrines. They certainly include an acceptance of the divinity of Christ and his Messiahship; though in what sense the reader will be more fully informed in later articles. Christian Scientists appear to lay little emphasis on rites, the outward forms of which they believe to be of small importance.

"Our baptism," says Mrs. Eddy, "is a purification from the flesh. Our church is built on Christ, the divine principle of the man Jesus. We can unite with this church only as we are new born of the Spirit—as we reach the life that is truth and the truth that is life—bringing forth the fruits of love, casting out error and healing the sick. Our eucharist is spiritual communion with the Father, the one Spirit. Our bread is that 'which cometh down from heaven.' Our cup is the cross-bearing inspiration of love—the cup that our Master drank, and of which he said, 'Drink ye all of it.' "

Christian Science churches have no pastors. The public services, usually held on Sunday morning and evening, are conducted by two readers, who must have had certain training in the principles of the faith. Frequently one of the two is a man, the other a woman. A morning service in the Chicago church is somewhat as follows: There are several hymns, selected from the hymnal published by the organization. Then one reader repeats the Lord's prayer, pausing after

each phrase to allow the other reader to add
Mrs. Eddy's "spiritual interpretation" thereof.
Whether this is an interpretation or not, one may
determine for himself by the following specimen:

> Our Father, who art in heaven,
>   "Our Father and Mother God, all harmonious."
> Hallowed be thy name,
>   "Adorable one."
> Thy kingdom come,
>   "Thy kingdom is come, good is ever present and
>   omnipotent."
> .   .   .   .
> And forgive us our debts as we forgive our debtors.
>   "And divine love is reflected in love."
> And lead us not into temptation, but deliver us from
>   evil.
>   "And leaveth us not in temptation, but delivereth
>   us from evil, sin, disease and death."

This is the only audible prayer used by the
Christian Scientists. They do not believe in
prayer in the ordinary Christian understanding of
that term. After this exercise follows the
responsive reading of a few sentences from the
scriptures and "Science and Health" by the readers and the congregation, using a printed leaflet
published by the denomination, in which the
services for all churches are the same for a given
Sunday. Then comes the "sermon," which is no
sermon at all, but a half-hour's reading by the
two readers, alternating, from the Bible and
"Science and Health," the selections being also
decided by the general committee of the church
at the Boston headquarters. An offering is fol-

lowed, not by a benediction, but by the reading of the "scientific statement of being," which, as a sort of brief creed, may be noted here:

"There is no life, truth, intelligence, or substance, in matter. All is infinite mind and its infinite manifestation, for God is all in all. Spirit is immortal truth; matter is mortal error. Spirit is the real and eternal; matter is the unreal and temporal. Spirit is God, and man is his image and likeness; hence man is spiritual and not material."

This closes the service. In the evening, the scripture selection is identical with the International Sunday-school lesson for the day. A midweek meeting is given to testimonies from persons who believe themselves to have been healed of disease by accepting the conception of the power of mind over matter which is a part of their religion.

THE CURES.

Intelligent Christian Scientists estimate the proportion of their number attracted to the subject and convinced by personal knowledge of physical cures as three-fourths of the whole. In other words, the remarkable book of Mrs. Eddy, whose obscurities, absurdities and self-contradictions must be evident to any educated reader not blinded by an unreasoning credulity, is not responsible for the wonderful growth of Christian Science. A few minds here and there, especially those easily mystified by abstract terms and the artful use of capital letters, have been converted solely through this book, without reference to its practical application. But the great majority have, either before becoming acquainted with "Science and Health," or during its study, learned of persons being suddenly and strangely released from the grasp of some chronic and severe disease while under the care of Christian Science practitioners. They have witnessed the swift transformation of an apparently helpless rheumatic or paralytic patient into an active and cheerful person. Or in their own bodies they have felt such changes, after proving for months or years the powerlessness of drugs to effect a radical cure. Let us be prepared to admit that great numbers of such cures have occurred. To attempt to break down the evidence on which they rest would be, in many cases, a waste of time; though certainly many persons have supposed themselves to be cured when they were really only temporarily relieved of pain; and many others have testified to wonderful cures of ailments which never existed save in their own imagination—such as certain cases of "nervous prostration," and similar ill-defined maladies which, any good physician knows, often need no cure save the firm resolution of the patient to throw away pills and powders, get out of doors, and take some interest in life. But cures have been wrought, and apparently in not a few cases where according to ordinary views of physiology the mind is not a controlling factor. This is the fact which Christian Scientists rely upon as their unassailable defense. To their minds it proves that disease is merely an illusion, an "error of mortal mind," that matter has no existence, that the whole fabric of crude philosophizing compounded by Mrs. Eddy from various ancient and modern idealists and from a method of scriptural exegesis more fantastic than that of Origen or the Talmud, is infallibly true. This is the most stupendous *non sequitur* in the history of delusions.

II

THE THEOLOGY OF CHRISTIAN SCIENCE

BY REV. H. H. BEACH

"Spiritual Science," "Mental Science," "Mind Cure," and similar cults—suckers on the thrifty stock of Christian Science, condemned and anathematized by the Boston people—are outside the scope of this paper.

Inasmuch as Christian Scientists frequently accuse their critics of misunderstanding their system and of garbling quotations from their writings, we are compelled both to prove our points by quotations and to quote whole sentences where only parts are relevant. We also preserve their capital initials, a harmless cipher by which the initiated distinguish the divine. The numbers refer to pages of the fifty-third edition of "Science and Health," unless otherwise explained.

SOURCES OF DOCTRINE.

Doubtless Mrs. Eddy is the source of Christian Science; though she claims to have found much of it in the Bible. She says: "In following these leadings of Scientific revelation, the Bible was my only text-book. The Scriptures were illumined, reason and revelation were reconciled, and afterwards the truth of Christian Science was demonstrated" (page 4). How thoroughly she searched the scriptures may be inferred from her comments, as, for instance, the following: "Jesus said of Lazarus: 'He is not dead, but sleepeth.' He restored Lazarus by the understanding that he had never died, not by an admission that his body had died and then lived again. Had Jesus believed that Lazarus had lived or died in his body, he would have stood on the same plane of belief with those who buried the body; and he could not therefore have resuscitated it" (page 241). We hardly need to refresh our recollection of John 11: 13, 14, to feel that there is a mistake somewhere: "Howbeit Jesus spake of his death, but they thought that he had spoken of taking rest in sleep. Then said Jesus unto them plainly: Lazarus is dead."

Like Swedenborg, Mrs. Eddy finds the spiritual sense of scripture by correspondences. Her correspondences are not the same as Swedenborg's, but equally wonderful. Adam is error; angels, messages, not messengers; burial, annihilation; Dan, animal magnetism; Gihon, the rights of woman acknowledged—morally, civilly, socially; Euphrates, Divine Science encompassing the universe and man; and Holy Spirit, Christian Science. New Jerusalem, also, is Divine Science. For a specimen, she reads, "Forgive us our debts as we forgive our debtors," as "Truth will destroy the claims of error" (page 322). Nor does she leave her pupils to wander about without a guide and lost in this Brobdingnag country, but gives them a "Key to the Scriptures," in which, at Genesis 2: 6, she says: "The continued account [of the creation] is mortal and material." "The history of error or matter, if veritable, would set aside the omnipotence of Spirit; but it is the false history in contradistinction from the true" (page 502). Such jugglery is a confession of judgment. The Bible does not teach Christian Science; Mrs. Eddy is the source; she admits that she has written all of it in one book, "Science and Health"; the whole Euphrates flows in this one channel. However, before we congratulate ourselves too heartily on finding all in one book, we should see the book. "Science and Health" may be scientific, but is hardly "knowledge duly arranged." Confusion, instead of humility, justifies the statement which its author somewhere makes, that Christian Scientists have "religious tenets" but not "doctrinal beliefs."

DOCTRINE OF GOD.

Begging Mohammed's pardon, there is no God but "Principle," "Mind," "Love," and Mrs. Eddy is its, his, her, prophet: "God is the Principle of Christian Science" (page 7). "The Principle of Divine Metaphysics is God" (page 5). "God is divine Principle, Supreme incorporeal Being, Mind, Spirit, Life, Truth, Love" (page 449).

"God is all" (page 7). The allness of I Cor.
15: 28—"that God may be all"—is order; but
Christian Science "allness" is substance and self-
hood. It blasphemously confuses finite souls with
God: "The term souls or spirits is as improper
as the term gods. Soul or Spirit signifies Deity,
and nothing else. There is no finite soul or
spirit" (page 450). "Divine Metaphysics, as
revealed to my understanding, shows me that all
is Mind, and that Mind is God, omnipotent, omnipresent, omniscient,—having all power, all presence, all Science" (page 171). "The Scriptures
say that God is all-in-all. From this it follows
that nothing possesses reality or existence except
Mind, God" (page 226). "The world believes in
many persons; but if God is personal, there is
but one person, because there is but one God"
(page 498).

Since Mrs. Eddy raises the question of personality, we press it: is her god a person? Here she
retreats into the mists of Hamilton's philosophy
of the Unconditioned and answers: "If the term
personality as applied to God, means infinite personality, then God is personal Being—in this
sense, but not in the lower sense" (page 10).
But even Mrs. Eddy's self-consciousness is too
strong for imagination and reverses the alter-
native: she assumes that we are persons, and this
god only a medical prescription. Hence a
patient, writing of Mrs. Eddy, says: "Her cures
are not the result of medicine, spiritualism or
mesmerism, but the application of a Principle

that she understands" (page 87). The gender of this supreme being is, on the whole, feminine: "We have not as much authority, in Divine Science, for considering God masculine, as we have for considering him feminine, for love imparts the highest idea of Deity" (page 498).

There is no personal trinity: "The theory of three persons in one God (that is, a personal Trinity, or Tri-unity) suggests heathen gods, rather than one ever-present I AM" (page 152). "Life, Truth and Love constitute the triune God, or triply divine Principle. They represent a Trinity in unity, three in one, the same in essence, though multiform in office: God the Father; Jesus the type of Sonship, Divine Science, or the Holy Comforter" (page 227).

The Christology of the Bible is the chief differentiating factor of Christian theology—"Hereby know ye the Spirit of God: every spirit which confesseth that Jesus Christ is come in the flesh is of God; and every spirit which confesseth not Jesus is not of God: and this is the spirit of the antichrist, whereof ye have heard that it cometh; and now it is in the world" (I John 4: 2, 3). As Ithuriel touched the toad "squat close to the ear of Eve," and Satan started up "discovered and expressed," so touch Christian Science with this verse of "celestial temper," and see what starts up. It appears in such statements as the following: "Wearing in part a human form (that is, as it seemed to mortal view), being conceived by a human mother, Jesus was the mediator between

Spirit and the flesh, between Truth and error"
(page 211). "Flesh" is "an illusion" (page 565).
It is too evident that Christian Science denies
Jesus Christ having come in flesh.

It may be seen that Jesus Christ was a Christian
Scientist, since Mrs. Eddy, referring to herself,
says: "So far as her knowledge of this matter
extends no other person has ever given Christian
Science to the world; but to those natural Christian Scientists, the ancient worthies, and to Jesus
the Christ, God certainly revealed its Spirit, if not
the absolute letter" (page 467). But he was not
as far advanced as he might have been, for,
again, referring to him, Mrs. Eddy says: "Had
wisdom characterized all his sayings, he would
not have prophesied his own death and therefore
hastened it." ("Miscellaneous Writings," six-
teenth edition, page 84.) He was, it is allowed,
begotten by the Holy Spirit, but the Holy Spirit
is Christian Science, and his conception was
natural: "Miracles are impossible in Science,
and here it takes issue with popular religions"
(page 249). "The illumination of Mary's spiritual
sense put to silence material law, and its order of
generation, and brought forth her child by the
revelation of Truth, demonstrating God as the
Father of men" (page 334). "The time cometh
when the spiritual origin of man, the Science
which ushered Jesus into human presence, will be
understood and demonstrated" (page 221).
"Until it is learned that generation rests on no
sexual basis, let marriage continue, and let us

permit no such disregard of law as may lead to a worse state of society than now exists" (page 274).

The burial of Jesus was his annihilation; he ceased forever to exist: "The Invisible Christ was incorporeal, whereas Jesus was a corporeal or bodily existence. This dual personality of the seen and the unseen, the Jesus and the Christ, continued until the Master's ascension; and then the human, the corporeal concept, or Jesus, disappeared; while the invisible, the spiritual idea, or the Christ, continued to exist in the eternal order of Divine Science, taking away the sins of the world, as the Christ had always done, even before the human Jesus was incarnate to mortal eyes" (page 229).

DOCTRINE OF MAN.

Mrs. Eddy distinguishes "man," "immortals" and "mortals" from each other.

"Man" has always been and will never cease to be, and beside him there is no other. He is the genus of which "immortals" and other divine ideas are species and individuals, the product of a final synthesis and the grandest achievement of realism. He is God's compound and infinite idea or reflection, and, like an individual "immortal," is, also, the mirror that reflects: "Separated from man, who expresses it, Spirit"—God—"would be a nonentity. Man divorced from Spirit would be equally a nonentity; but there is, there can be, no such division, for man is co-existent with God" (page 461). In short, this "man" is to this god as a construction company to a railroad—identical for profits, but distinct for losses.

"Immortals" are the individual ideas, mirrors and reflections of the god. The "reflection" notion seems to be a misuse of the "brightness" of Heb. 1 : 3. Every "immortal" has always been and will never cease to be, and is absolutely perfect. They are the triumph of idealism. On the one hand they are the stuff that "man" is made of, and on the other the doubles of "mortals."

"Mortals," or "mortal mind," in the nomenclature of Christian Science, are simply human beings, as we know ourselves and others—body, soul and spirit: "Mortals will disappear, and immortals, or the children of God, will appear as the only and eternal verities of man. Mortals are not fallen children of God. They never had a perfect state of Being, which may subsequently be regained. They were from the beginning of mortal history, conceived in sin and brought forth in iniquity. Mortals are material falsities. In the words of Paul, they are without hope and without God in the world. They are errors, made up of sin, sickness and death, which must disappear to give place to the facts which belong to immortal man" (page 460). "Temporal life is a false sense of existence" (page 16). "Think of thyself as the orange just eaten, of which only the pleasant idea is left" (page 257).

Unlike Frankenstein's creature, all of Mrs. Eddy's men and women have good characters—or no characters: "Sin and mortality" are "native nothingness" (page 177). "A wicked man is little else than a creature of error" (page 185).

This anthropology is materialism reversed but not converted, for to say that matter is mental is practically the same as to say that mind is material: "Electricity is not a vital fluid, but the least material form of human consciousness—the material thought essence, which forms the link between matter and mortal mind" (page 189). Matter is the "lower substratum" of mortal mind (page 93). "The fading forms of matter are the fleeting thoughts of the human mind" (page 160). "Nothing we can say or believe regarding matter is true, except that matter is unreal, and is therefore a belief, which has its beginning and ending" (page 173).

As a cap-sheaf tenet, we are all insane: "There is a universal insanity, which mistakes fable for fact throughout the entire round of the material senses; but this general craze cannot, in a spiritual diagnosis, shield the individual case from the special name of insanity. Those unfortunate people who are committed to insane asylums are only so many well-defined instances of the baneful effects of illusion on mortal minds and bodies" (page 406). It is hardly to be wondered at.

As the limits of this paper preclude extended comments, we must trust that these vices of the mind are monsters of "so frightful mien as to be hated" need "but to be seen"; but we cannot

refrain, here, from calling attention to a fatal gap
in the system. It lies between "mortals" and
"immortals." "Mortals," though, like Iago's
purse, "something, nothing," are bad, and so
cannot have sprung from the "immortals," or
from the "man" or from the god. This is a
fundamental tenet: "It is self-evident that this
Mind, or Divine Principle, can produce nothing
unlike Itself—Himself, Herself" (thirtieth edition). What has produced mortals? In what
do they inhere? Perhaps it is a case of spon-
taneous combustion. There is neither source nor
support for them: "The immortal never produces
the mortal, and Good cannot result in evil"
(page 173). The earth rests on the elephant, and
the elephant stands on the turtle. Mrs. Eddy
may go on shoring up the system, but when she
stops the whole structure falls in ruins.

DOCTRINE OF SALVATION.

Full Christian Science salvation is a repentance
from the nothingness of mortal beliefs to the
somethingness of divine ideas. Jesus attained it.
When his disciples thought his body was lying
pale and cold in Joseph's tomb, he was only hid-
ing there, studying Christian Science: "The
lonely precincts of the tomb gave Jesus a refuge
from his foes, and a place in which to solve the
great problem of Being" (page 349). Every one
will, eventually, reach it: "The dream that
matter and error are something, must yield to
reason and revelation. Then mortals will behold

its nothingness, and sickness and sin will disappear to their vision" (page 293).

But neither salvation nor anything else is effected by supplicatory prayer: "God is not influenced by man" (page 313). "Who would stand before a blackboard and pray the principle of mathematics to work out the problem? The rule is already established, and it is our task to work out the solution. Shall we ask the divine Principle of all goodness to do his own work? That work was finished long ago; and we have only to avail ourselves of God's rule, in order to receive the blessing" (page 308). Nor is one saved by a vicarious atonement: "Final deliverance from error—whereby we rejoice in immortality, boundless freedom and sinless sense—is neither reached through paths of flowers, nor by pinning one's faith to vicarious effort" (page 499, thirtieth edition). The means of salvation are suffering and science, church life and metaphysical practice: "Either here or hereafter, suffering or Science must purge man of false illusions about life and mind, and cleanse him of material sense and self" (page 482, thirtieth edition). "The followers of Christ must drink his cup for centuries to come" (page 483, thirtieth edition). "Through discernment of the spiritual opposite of materiality, even the way through Christ, Truth, man will reopen with the Key of Science, the gates of Paradise, which human beliefs have closed, and will find himself unfallen, upright, pure and free, not needing to consult almanacs for the probabilities of Life, or to study brain-ology in order to learn how much of a man he is" (page 63).

But five dollars' worth of metaphysical practice is better than a one-hundred dollar order on an idealized "immortal," and healing is the principal part of salvation. Indeed, one might think it the whole, for Mrs. Eddy says: "The term Christian Science was introduced by the author to designate the Scientific system of Metaphysical Healing" (page 17). The instruments of this spiritual surgery are thought-suggestion and will-pressure. When the practitioner invades the sick man's common-sense, either honestly through the gate, by spoken word, or burglariously over the wall, by silent, present or absent treatment, the resultant trouble is called "chemicalization": "Disease is a fear, expressed not so much by the lips as in the functions of the body" (page 372). "The author never knew a patient who did not recover when the fear of the disease was gone" (page 376). "Destroy fear, and you end the fever" (page 375). "Anodynes, counter-irritants, and depletion never reduce inflammation; but the Truth of Being, whispered into the ear of mortal mind, will bring relief" (page 373). "Explain audibly to your patients (as soon as they can bear it) the utter control which Mind holds over the body" (page 415). "If the case is that of a young child or an infant, it needs to be met mainly through the parents, silently or audibly on the strictest rules of Christian Science" (page 411). "It is more difficult to make yourself heard mentally when others are thinking about your patients, or conversing with them" (page 422). "Like Jesus, the healer should speak to disease as one having authority over it, leaving Soul to master the false evidences of the corporeal senses, and assert its claims over mortality and sickness" (page 393). Mrs. Eddy appeals to the recovery of the sick to prove that Christian Science is true, but unfortunately admits that "the symptoms of disease" may be relieved by deception and error: "Homœopathic remedies, sometimes not containing a particle of medicine, are known to relieve the symptoms of disease" (page 397).

LAST THINGS.

Natural death is but a mortal illusion, and may be followed by an intermediate state of illusion like the earthly condition. The coming of Christ is an awakening from the mortal dream that sin and sickness are realities; but neither Christ nor other "Immortal" ever returns to earth. There is no final judgment, and God will never punish sinners. The saved condition, the grand consummation, is a sublime lonesomeness: "Would existence be to you a blank without personal friends? Then the time cometh when you will be solitary, left without sympathy and alone; but the human vacuum is to be filled with divine Love" (page 162).

The theology of Christian Science, taken as a whole, is doubtless a kind of pantheism. The mortal-immortal-man-goddess is all. But what kind of a creature is she? We discern her chief features clearly enough. Disclaiming ridicule, that which is specially the god is the intelligence; "man" is the whole form—muscles, tissues, integuments; "immortals" are innumerable arms, reaching out forever into undiscovered spaces; and "mortals" are grotesque shadows of all that is tangible, specters of Brocken, on a background of mist.

III

THE PHILOSOPHY OF CHRISTIAN SCIENCE

BY PRES. WILLIAM H. P. FAUNCE, D.D.

All persons familiar with the intellectual life of our time are conscious of a wave of "new thought" now sweeping over this country. This thought assumes Protean forms, and manifests itself in a mass of literature of all shades from the sublime to the ridiculous. The movement has a two-fold origin. On the one hand, it comes from the German idealism of Hegel and Fichte, which (mediated by Thomas Hill Green) has at last filtered down through all the strata of society and reached the average man. On the other hand, it comes from contact with the religions of the Orient, and a new appreciation of their mystic peace and brooding calm.

A foretaste of this "new thought" appeared in the New England Transcendentalism of fifty years ago; it achieved its brightest literary expression in Emerson, and its passing embodiment in the Brook Farm experiment. But that movement was chiefly confined to New England. The present movement—a reaction from the deistic view of the world which has long pervaded

both science and theology—covers the entire
country, and is putting forth a quantity of literature of whose extent few are aware. The philosophy underlying the whole is optimistic and
idealistic, and often claims and produces large
results in bodily healing. Mrs. Eddy is only one
—the most successful one—of scores of teachers
in this country who are now insisting on the
power of thought to change life, and the immanence of God in such a sense that pain and grief
and sin can be practically ignored.

A CRUDE IDEALISM.

Most of these teachers are destitute of philosophical training, and are putting forth crude
systems more wonderful than Joseph's coat which
was *not* "of many colors." "They have been at
a great feast of languages and stolen the scraps."
They strongly antagonize each other, and unite
only in antagonizing both materialism and
scholastic orthodoxy. Oriental importations, the
flotsam and jetsam of the Parliament of Religions,
wander through the country, unfolding outworn
theories of the Orient as the latest fad of the
Occident. Indian Swamis enter Boston parlors
and instruct companies of adoring women in the
science of mist and moonbeams. Some of Mrs.
Eddy's pupils, weary of her personal control,
have revolted and set up schools of their own.
"Metaphysical healing" is largely practiced in the
eastern states by those who utterly reject Christian Science. On a much higher intellectual level

are the books of Dr. Dresser, Ralph Waldo Trine and Henry Wood, all having an extraordinary sale, all insisting that "there is nothing either good or bad but thinking makes it so," all giving an idealistic and spiritual interpretation of the universe, and all succeeding in lifting from scores of weary souls a burden of care and fear and pain which we have been taught is inalienable from human life. All of these teachers unite in rejecting the eighteenth-century conception of God as an "absentee," or as an "occasional visitor," or as a "magnified Lord Shaftesbury"; and when they are theists in any real sense, affirm that God is immanent in the human soul, and that if we will but "practice his presence" we shall be delivered from all the ills consequent on faith in a distant deity. We may at least rejoice that the tendencies of our time are no longer toward disbelief in a spiritual world. So far has the pendulum swung, that the same popular literature which, thirty years ago, was trying to believe that "thought is a secretion of the brain," now gravely affirms that the brain is a figment of thought!

What, now, is the particular phase of this thought embodied in "Christian Science"? The history of the movement has been given in a previous article. We are concerned now only with its philosophy, which is most certainly a rare collection of shreds and patches. Among the many notions inconsistently united we may distinguish a few dominant thoughts.

1. The idealistic conception of matter. A modern teacher has called Christian Science "an incomplete misconception of Berkeleyanism." But good Bishop Berkeley's faith in Christianity was not hindered in the least by his philosophic explanation of the material universe. In order to combat atheism and materialism, he asserted and believed that the apparently external world exists only in our own idea, and that minds alone have real existence This is a philosophic view which will always have its advocates, and need not be discussed here. All human beings must *act as if matter did exist*, and the speculative denial has little influence on life. If the Christian Scientist wishes to build a house, he must treat bricks, mortar and timber just as every other man treats them, even though he honestly believes that the bricks are all in his own brain. Only in the treatment of the human body does the idealism have practical consequences. If a man believes that his body is the pure expression and even the creation of mind, he will certainly endeavor to shape and control that body mainly through the maintenance of mental conditions.

2. A monistic conception of God and his world, ever verging into pantheism. The publication of Dr. Strong's "Ethical Monism" was one of the most significant events in the history of Christian thought in this country. It showed how great is the present reaction from a mechanical and external theology, and how even the thinkers who have stood most stoutly for the reality of the soul,

of sin, and of redemption, are now passionately demanding some unifying conception of the world-order. Dr. Strong is abundantly able to safeguard his monism; but Mrs. Eddy goes straight over into the camp of those who deny personality to God, and all real freedom and responsibility to man. She explicitly denies that God is personal. Her denial is meant as a protest against anthropomorphism; but it goes so far as to make God a metaphysical abstraction or principle.

DENIES DIVINE PERSONALITY.

A young man recently came to me who had gone through Christian Science into Atheism. I asked him to describe the path he had passed over. He answered: "The Christian Science teacher began by thoroughly persuading me that God is not personal, but is pure 'Principle.' After some months I accepted that, and then I said to myself: 'What is a principle? Does it have real existence? Is it an entity or reality?' I soon saw that a 'principle' is simply an idea of my own mind, and when the Scientist dissolved my God into 'principle' I ceased to believe in any God whatever. I now believe simply in myself."

Mrs. Eddy answers the question, "What is God?" as follows: "God is divine Principle, supreme incorporeal Being, Mind, Spirit, Soul, Truth, Love." At the head of this confusing list of alleged synonyms she puts Principle, as being the most thoroughly de-personalized term, and hence best suited to her purpose. But let us ask her to define more closely. Does her God possess consciousness, will, purpose? Is he so like to the noblest earthly father that our highest name for him is Heavenly Father, and that we can commune with him, pray to him? To all this Mrs. Eddy must answer, No. To her God one must not pray, for that would be to acknowledge him as personal. While one might in an unwary moment call her God "Father," yet that term is omitted from her definition of God. Her God is "Being," but being need not be conscious of its own existence, or of ours. Her God is Truth; but truth is destitute of volition or affection for man. Her God is Life; but life in moss and tree is unconscious and unintelligent. Her God is Love; but not the love which can answer any request for aid. Her God is Mind, Spirit, Soul, provided that we interpret those words as synonyms of unconscious "principle."

ELASTIC TERMS.

Much of the success of Christian Science is due to the fact that its vague phraseology is equally acceptable to the evangelical Christian and to the atheist. The average Christian, approaching the Christian Science creed on one side, hears that God is "Spirit, omnipresent and eternal," and at once accepts the teaching. The atheist, coming up on the other side, hears that the only God is "principle, truth, harmony," and he can accept it without the slightest change of position. I would not charge conscious duplicity upon Christian Science teachers. But I do know that they will say to the simple-minded Christian: "We of course believe in prayer, and we use the Lord's prayer at every service;" while to the antagonist of Christianity they will say: "You know in what sense we believe in prayer—it is by affirming Principle."

When Mrs. Eddy in her reaction from deism joins Theodore Parker in denying personality to God, she makes her capital blunder out of which all other blunders spring. She thinks personality means limitation and corporeality. But Dr. Martineau has well said: "You cannot deny God's personality without sacrificing his infinitude: for there is a mode of action—the *preferential*—the very mode which distinguishes rational beings—from which you exclude him." Since Mrs. Eddy's deity is incapable of preferring and willing and seeking moral ends and communing with his children, since he is *less than personal*, he is less than the Christian God, the Father of our Lord Jesus Christ.

3. Of course such a faith must issue in optimism. Pain vanishes, since God is incapable of pain, and God is the only reality. Mrs. Eddy is as contemptuous toward pain as was Marcus Aurelius when he wrote: "Do not suppose you are hurt and your complaint ceases. Cease your complaint and you are not hurt." Indeed, her steadfast denial, i. e., her determination to ignore, has close affinity with ancient stoicism. There is nothing new under the sun. The

Emersonian oracle has long been telling us that "good is positive, evil only privative"; Browning has long been crying, "All's right with the world." But what the stoics and the poets have always affirmed as ideally true, Christian Science turns into bald prose propositions, telling us that, by refusing to think of the ills that flesh is heir to, we may destroy their phantasmagorial existence.

Still further goes this optimism. It denies that sin exists, save in our thought of it, i. e., in "mortal mind." It declares that "man is incapable of sin, sickness and death, inasmuch as he derives his essence from God, and possesses not a single original, or underived, power." Here again we are misled by alleged synonyms. Certainly man has no "underived" power; but has he no "original" power? Has he no power to originate action, to determine some events, to choose between good and evil? If not, we are landed at once in the pantheism where good and evil coalesce in universal being. The Christian church has always believed in a God not to be identified with his own creation, a God distinct, though not separate from his children, a God with power

"To create man, and then leave him
Able, his own word saith, to grieve him."

Mrs. Eddy denies that man is able to grieve God, both because God is incapable of grief or any other emotion, and because all human sin is apparent only, and in reality does not exist. Such teaching is exceedingly perilous to the moral life. It coincides with the teaching of the English Antinomians of the seventeenth century, who affirmed that "the feelings of conscience, which tell them that sin is theirs, arise from a want of knowing the truth." In the view of Christian Science, since man is incapable of sin, conviction of sin is a dream, and redemption from it an impossibility. Christ therefore is the "Way-shower," no longer himself the Way. When a leading Christian Scientist said to me: "Mrs. Eddy is the way to God," I answered, "I thought Christ was the Way." "But Christ, you know, is dead," she answered, "and Mrs. Eddy is now alive." "But Mrs. Eddy must soon die, and who then will be the way?" "Well, we do not think that Mrs. Eddy will—what you call *die;* we expect she will—dissolve—into—the life of the universe!" Could optimism further go? Yet men call this age—the age of Keely and Mrs. Eddy—a materialistic age!

TWO TRUTHS AFFIRMED.

Let me now mention briefly some of the strong and the weak elements in this strange Christian Science creed.

It is strong in its clear realization of the immanence of God. God is not only "in his heaven," but God is in his world. The average Christian church is still shy of the doctrine of the Holy Spirit, leaving that to Northfield and Keswick, and believes only in a far-away deity who occasionally has interfered with his world to work a miracle. The average Christian church believes in an inspiration which ceased about 100 A. D., and miracles which ceased about 300 A. D., and in an interpretation of the Bible which makes it the story of what was but no longer is. Christian Science affirms that God is as near his world to-day as in any age, and performing all the wonders now that he ever performed. In this it agrees with the Roman Catholic Church, and is a standing reproach to Protestant unbelief.

Christian Science has undoubtedly gotten hold of a great truth in its affirmation that the best way to heal the body is through the mind. The principle which underlies all these various forms of healing, "metaphysical," "mental," "faith," or "Christian Science" healing, is the same, as Dr. Buckley has clearly shown, or Mr. Hudson in his "Law of Psychic Phenomena." If we believe that the mind is more than the body, and that all our minds are in contact with the infinite Mind, why should we not, when afflicted with bodily disorder, go first, not to the druggist, but to some friend of strong mental and spiritual powers? If we believe in prayer, why should we not pray to that Spirit in which we "live and move and have our being"? Medicine has long treated the mind through the body; now let it show as much zeal in treating the body through the mind. It is for this reason that Prof. James, of Harvard University, has recently defended the Christian Scientists against the enaction of an oppressive law by the Massachusetts Legislature—not because he can accept their bizarre philosophy, but because he believes that the power of the mental over the physical life is greater than any accredited philosophy has been willing to admit, and that the possession of a medical diploma does not entitle any man to a monopoly of healing.

The weakness and danger of Christian Science are to be found, especially, in the following points:

1. In a quite unwarranted use of the Bible. Mrs. Eddy professes greatest attachment to the scriptures, and her followers are constant readers of the Bible. Yet she selects only certain portions of the Bible, and commends those portions only when interpreted allegorically and arbitrarily. Thus, in commenting on Gen. 2: 7, she mildly queries: "Is it the truth? or is it a lie, concerning man and God? It must be the latter, for God presently curses the ground." Mr. Ingersoll, in elucidating the "Mistakes of Moses," never condescended to such language. But not content with the charge of falsehood, she proceeds to a little exegesis of her own. In order to prove that "Adam" is merely a name for the "matter" which opposes "mind," she suggests to her obedient followers a short and easy method with the skeptics: "Divide the name Adam into two syllables, and it reads, *a dam* or obstruction." And the book which contains this sample of exegesis is supposed to be addressed to sane men and women!

The Bible flames from beginning to end with a passion for righteousness, and an indignation against iniquity. To say that man is "incapable of sin" is to stultify the noble army of martyrs, to discredit all the prophets and apostles, and to make the life and death of Christ farcical. Men may believe that sin is temporary, that at last God shall be all in all, and still follow the Christ. But to say that man is "incapable of sin" is to rob man of real freedom and responsibility, and make Christ only a "Way-shower" instead of the Way.

2. Another danger lies in a dissolving of God into a misty, unconscious abstraction. In her attempt to get rid of anthropomorphism Mrs. Eddy denies personality. But John Fiske has truly said: "We are bound to conceive of the Eternal Reality in terms of the only reality that we know, or else refrain from conceiving it under any form whatever." To ascribe human weakness and limitation to God, is indeed an error and a folly. But all the objects we know are either persons or things. Which is Mrs. Eddy's God? Does she believe in the thingness of God, or the *personality* of God? The answer of her books is clear—she believes in the *thingness* of God, in a Substance like that of Spinoza, incapable of purpose, choice, or consciousness, a Being whose shadowy self is best described as "Principle."

MONEY IS NOT DESPISED.

3. A danger in this movement, which perhaps has not yet developed, is the confusing of moral

distinctions through the denial of the reality of
evil   I gladly bear witness to the personal
nobility and high-mindedness of many Christian
Scientists.   I rejoice to find beauty in lives
whose creed I cannot accept.   Yet the Scientists
are often sorely put to it to explain how Mrs.
Eddy could charge each pupil $300 for twelve
lessons, or what she does with the vast revenues
which constantly come into her hands.   The
usual explanation is that any sum is small in
comparison with the benefits received, and that
"all that a man hath will he give for his life."   In
absolute religious despotism, combined with
belief that one is "incapable of sin," however we
may explain the words, danger always lurks.
The latest step in this line, and perhaps the most
surprising, is the publication in the Christian
Science Weekly for January 19 of the following
card concerning "Christian Science Souvenir
Spoons": "On each of these most beautiful
spoons is a motto in bas-relief, that every person
on earth needs to hold in thought.   Mother
requests that Christian Scientists shall not ask to
be informed what this motto is, but each Scientist
shall purchase at least one spoon, and those who
can afford it, one dozen spoons, that their families
may read this motto at every meal, and their
guests be made partakers of its simple truth.
(Signed).   Mary Baker Eddy."   Probably one
outside the mystic circle should not "ask to be
informed" as to the price of these precious
spoons; nor as to the object of the sale; nor as to

the proceeds of the sale, if every Scientist of the
300,000 claimed in this country were to purchase
one or one dozen; nor as to the results if "every
person on earth" should seek after this talismanic
motto, nor whether the "Christian Science
Souvenir Company" is identical with the "Church
of Christ, Scientist." But even one outside the
circle may think, and marvel, and wonder if all
the followers of the "Mother" will approve.
"Beloved, believe not every spirit, but try the
spirits whether they are of God."

IV

THE INHERENT DIFFICULTIES OF
CHRISTIAN SCIENCE

BY J. W. CONLEY, D.D.

The one who undertakes to point out the inherent difficulties of Christian Science is likely to be told that he has entirely failed in understanding the subject discussed. His unspiritual and material view of things has blinded him to the merits of the case, and has led him to see difficulties where none really exist. But Christian Science must meet the difficulties in the minds of thoughtful, intelligent people in some other way than by charging lack of mental acumen or of spiritual apprehension. However difficult it may be to get at the meaning of many of Mrs. Eddy's ambiguous statements, still it is not a very difficult matter to understand the principles involved in her teachings. Many who have been favorably impressed by Christian Science through its more attractive features have never seriously considered the inherent difficulties of the system.

1. The infallibility of Mrs. Eddy may well claim attention first. While there are several new schools of Christian Science which do not accept the leadership of Mrs. Eddy, these are not

properly Christian Science. If Mohammedanism
should give up Mohammed and the Koran, it
might still retain some features of the old system,
but it would cease to be Mohammedanism. Mrs.
Eddy claims that to her alone in modern times
has been given the revelation of the Truth. She
says: "God had been graciously fitting me, during many years, for the reception of a final
revelation of the absolute Principle of Scientific
Mind-healing." The book "Science and Health"
is a setting forth of this "final revelation" of the
"Absolute Principle." To her has been made
known the true gospel which has been hidden for
centuries. In her address, read at the dedication
of the First Christian Science Church in Chicago
in 1897, she says: "It is authentically said that an
expositor of the dates of Daniel fixed the year
1866 or 1867 for the return of Christ—the return
of the Spiritual Idea to the material earth or
antipode of heaven. It is a marked coincidence
that those dates were the first two years of my
discovery of Christian Science." Here she evidently claims that she represents the second coming of Christ. The way in which her followers
regard her is set forth in the following statement
explanatory of the fact that instead of having
preaching in their services, they have extracts
read from Mrs. Eddy's writings: "The canonical
writings, together with the word of our text-book,
corroborating and explaining the Bible texts in
their spiritual import and application to all ages,
past, present and future, constitute a sermon

undivorced from truth, uncontaminated and
unfettered by human hypotheses, and *authorized*
by Christ." Her infallibility is of a most extraordinary character. She speaks the final word
upon the whole range of human thought. She
has discovered the Final and Absolute Principle.
There can be nothing beyond. To depart from
her teachings is, to use her own words, to "adopt
and adhere to some particular system of *human*
thought." The general acceptance of such an
infallibility would stop all independent thinking,
and put an end to all human progress.

2. A second difficulty is the denial of the testimony of the senses and of the reality of matter.
Mrs. Eddy declares, "Nothing we can say or
believe regarding matter is true except that
matter is unreal." She defines matter as "that
which mortal mind sees, feels, hears, tastes, and
smells only in belief." "[Christian] Science
reveals material man as a dream at all times, and
never as the real Being." "The material senses
testify falsely." "The material atom is an out-
lined falsity of consciousness." "[Christian]
Science and material sense conflict at all points
from the revolution of the earth to the fall of a
sparrow." Scores of similar quotations might
be given. The whole material universe, from
remotest star to minutest atom, is a strange, inex-
plicable, complicated delusion. It is the product
of the "mortal mind," and the "mortal mind" itself is defined as a "solecism in language," "something untrue and unreal." Sun, moon, and stars,

changing seasons, waving forests, broad rivers, majestic mountains, verdant landscapes, blooming flowers, singing birds, ripening harvests, are all a delusion. The psalmist was blind and misled when he wrote, "The heavens declare the glory of God." And again, when he said, "When I consider thy heavens, the work of thy fingers, the moon and the stars which thou hast made," he was like a man talking in his sleep. This attitude toward matter and the testimony of the senses is more serious than at first it may appear. It means that it is folly to study astronomy, geology, chemistry, physiology, biology or any of the sciences which have contributed so marvelously to the progress of this age. Such studies only tend to more hopelessly involve man in the bondage to that which is fundamentally false. Christian Science is really a call to close all our laboratories and all schools of scientific research. Already in a number of instances, Christian Scientists have requested that their children be excused from studying physiology in the public schools. This is in entire harmony with their faith. But the same principle that calls for the rejection of physiology will, when fully and consistently applied, reject all scientific study. Mrs. Eddy says: "If half the attention given to hygiene were given to the study of Christian Science and its elevation of thought, this alone would usher in the millennium."

But this rejection of the testimony of the senses means universal skepticism. The one indivisible mind communicates with the material world and also with the spiritual. On the one hand it affirms the reality of objects and things, on the other of thoughts and ideas. If all is a delusion on one side, what possible ground have we for believing in the affirmations of the same mind on the other side? The logical conclusion, from the rejection of the testimony of the mind respecting the senses, is to reject all its testimony and to believe that nothing is real, and that existence itself is

A DREAM WITHOUT A DREAMER.

3. Several serious difficulties present themselves in connection with the healing of disease. It is urged that all suffering originates in the mind. There is no objective disorder, because there is no object to be disordered. The entire conception of material existence is that it is evil and evil only. It is regarded as a delusion, but it is the reverse of good. Mrs. Eddy declares: "The only conscious existence in the flesh is error of some sort, sin, pain, death." Note carefully this statement—"the *only conscious* existence in the flesh" is utterly bad—"sin, pain, death."

Here is a pessimism equal to that held by Schopenhauer. Everything in the material world is the direct opposite of good. Humanity is one great disease. It is a contradiction of terms to speak of physical health, for the body itself is wholly and irremediably evil. You can-

not cure the body, for the body is incurable; it is
from its very nature "sin, disease, death." We
are told that the way to get rid of a pain in the
head is to insist that there is no such thing as
pain. Why not be consistent and insist that there
is no such thing as the head? The real trouble is
not that we have supposed that there is a pain,
but that we have been deluded into believing
that there is a head to have a pain. Here is
a fundamental inconsistency in the Christian
Science treatment of disease. Disease is simply
a symptom. Why treat symptoms? Why not
think truth respecting the body, and cast off at
once this wretched pest-house of pain, this
disease-infected delusion? The difficulty is not
that there is a pain here, and a pain there, but
that there is a material body at all. Here is the
root of the whole trouble. The weakest point
logically in Christian Science is right here where
it claims to be strongest. It does not apply its
fundamental principle in treating disease. Its
cures must find some explanation aside from the
application of a principle which is never really
applied.

A somewhat extended quotation will prepare
the way for another difficulty:

"If a dose of poison is swallowed through mistake and the patient dies, even though physician
and patient are expecting favorable results, does
belief, you ask, cause this death? Even so, and
as directly as if the poison had been intentionally
taken. In such cases a few persons believe the

potion swallowed by the patient to be harmless, but the vast majority of mankind, though they know nothing of this particular case and this particular person, believe the arsenic, the strychnine, or whatever the drug used, to be poisonous, for it has been set down as poison by mortal mind. The consequence is that the result is controlled by the majority opinion outside and not by the infinitesimal minority of opinion in the sick chamber."

And yet elsewhere in direct contradiction to this she declares: "Homeopathic remedies, sometimes not containing a particle of medicine, are known to relieve symptoms of disease. What produces the change? It is the faith of 'mortal mind.'" But here evidently it is the minority inside and not the vast majority outside that does the work. If the minority could make helpful medicine out of unmedicated pills, why could not a minority make poison harmless? There must be something wrong here somewhere. Then, too, there are many mind cures, hypnotic cures, spiritualistic cures, and the like. These Mrs. Eddy maintains are effected by the belief or power of "mortal mind." Here also it would seem that the minority has in some unaccountable way caught the majority napping.

WHY NOT HEAL ALL DISEASES?

Still another difficulty is met with in connection with the work of healing. Why do some diseases yield to Christian Science treatment much more readily than others? A careful statement of the

principle upon which these cures are based will
show the inconsistency here. "Immortal Mind"
is universal and real, as "mortal mind" is limited
and unreal. To cure disease "Immortal Mind"
must assert the unreality of the affirmations of
"mortal mind." "Mortal mind" says, "I am sick."
"Immortal Mind" says, "There is no such thing
as sickness, it is all a delusion," and since
"Immortal Mind" is infinitely superior to
"mortal mind," the supposed sickness vanishes.
Such being the case, it follows that healing is not
in any way due to faith in the divine mercy and
power, nor is it a question of magnetism, hypnotism, or the power of mind over matter; it is
simply accepting and thinking the truth. This
being true, there is no place whatever for any
distinction as to diseases; one ought to yield as
readily as another. When Copernicus discovered
the right principle of the universe the big planets
fell into line just as readily as the little ones. If
Christian Science is all it claims to be, it ought to
be able to set a broken bone and restore lost
teeth, just as readily as to cure a nervous head-
ache. Again, if Christian Science has indeed discovered the great underlying principle for the
banishment of disease, its work of healing ought
to show a marked superiority over every other
method for the treatment of the suffering. But
the fact is that all of its works are readily paral-
leled by other schemes for healing without
medicine, and seem to have the same character-
istics and limitations that these have.

4. Another serious difficulty is the origin of evil. The simple fundamental principles are as follows: "God is all, God is good, hence all is good." There is absolutely nothing outside of God or good. And yet somewhere, somehow, a vast and awful delusion sprang up which has filled the world with groans and sighs and sins. In commenting upon Gen. 3: 4, 5, Mrs. Eddy says: "This myth represents error as always asserting its superiority over truth, giving the lie to Divine Science, and saying through the material senses, 'I can open your eyes, I can do what God has not done for you.'" One writer has very pertinently said, "While Mrs. Eddy smuggles in the fact of the fall, she gives no rational account of it." She attempts to explain it in the following: "The history of error is a dream-narrative. The dream has no reality, no intelligence, no mind; therefore the dreamer and the dream are one, for neither is true or real." But if we grant this, the difficulty is not met. The dream itself is a fact, a horrible, protracted nightmare. "God is all, and all is good;" whence then this awful dream of evil? Man is not responsible for it, for he is simply an idea of the Divine Mind. He is a part of God, of the universal good. He has no independent personality or power. Call evil a dream, a delusion, a fancy, anything you please, still it remains a fearful fact in human experience. It may be urged that other systems of thought have serious trouble in accounting for evil. But the difference is this,

they adopt principles that recognize the possibility and power of evil, while Christian Science by its fundamental principles excludes all possibility of evil in any form whatever. There is no place logically in this system for even a dream of evil.

THE TREATMENT OF SCRIPTURE.

5. The treatment of scripture is another difficulty of large proportions. Mrs. Eddy claims to be a firm believer in the Bible. She declares "Divine Science derives its sanction from the Bible." "The Bible has been my only text-book. I have no other guide in the strait and narrow way of this science." She makes frequent quotations from the scriptures, not always accurate, however, as the following example will show: "Jesus said of Lazarus, 'He is not dead, but sleepeth.' He restored Lazarus by the understanding that he never died, not by an admission that his body had died and then lived again." Now the fact is, Jesus never used the above language of Lazarus. He said first, "Our friend Lazarus sleepeth;" but when the disciples misunderstood him he said plainly unto them, "Lazarus is dead."

Her interpretations of scripture have no regard whatever for the ordinary meaning of words or for the generally accepted laws of language. The comment on Gen. 1: 20 is a fair sample of her method. "To mortal mind the universe is liquid, solid and aëriform. Spiritually interpreted, rocks and mountains stand for the solid and grand ideas of truth. Animals and mortals metaphorically present the gradations of thought, rising in the scale of intelligence taking form in masculine and feminine ideas. The fowls which fly above the earth, in open firmament of heaven, correspond to aspirations soaring beyond and above corporeality, to the understanding of their incorporeal and divine Principle." Such handling of scripture undermines all sane use of language, and makes the Bible capable of teaching any vagary that may enter the mind of man. But worse than this, she really denies all the great doctrines of the Bible. Resurrection means "Spiritualization of thought." "Holy Ghost means Divine Science." "Baptism is purification by Spirit, submergence in truth." "Angels are God's thoughts passing to man." "Truth bestows no pardon upon error." "We cannot escape the penalty due for sin." "The theory of three persons in one God suggests heathen gods rather than the one ever-present I AM." "Life, Truth, and Love constitute the triune God or triply divine Principle." "Every mortal must learn that there is no power in evil." And thus she goes on until the whole gospel structure is utterly undermined.

There is space simply to mention other grave difficulties. Mrs. Eddy constantly misrepresents current philosophic and scientific thought respecting the nature of matter. No one believes that matter is "a creator" or is in itself "sentient." A system that makes all sin an illusion seems to undermine all moral distinctions. There has not yet been time to see what the fruitage in this respect will be.

The personality of God and also of man as universally understood by Christendom is denied. God is Principle and not person, and man is an idea of God, and Jesus Christ was simply "the highest human corporeal concept of the divine idea." His divinity was essentially the same as that of every man. Christian Science dethrones the divine Christ. There is no place left for the exercise of the mercy of a loving personal Father. An absolute all-embracing Principle is supreme. Of necessity, this system becomes practically prayerless. Mrs. Eddy declares: "God is not influenced by man." She argues against audible prayer: "We must close the lips and silence the material senses." In fact, there is no place for what is ordinarily understood by silent prayer. "The habitual struggle to be always good is unceasing prayer." And she adds: "He who is immutably right will do right without being reminded of his promises." All this is directly contrary to the frequent Biblical examples of and exhortations to prayer.

Christian Science is a strange mixture of theosophy, idealism, pantheism and Swedenborgianism, with a thin veneering of professed science and Christianity, and must certainly be greatly modified or utterly break down under the weight of its inherent and insuperable difficulties.

V

EXPLANATIONS OF THE GROWTH OF CHRISTIAN SCIENCE

BY LATHAN A. CRANDALL, D.D.

In every discussion of Christian Science this question is sure to be asked: "How do you account for its growth?" It is not necessary to repeat the statistics presented by Mr. Slater in the first article of this series, in order to show that this question is a most natural one. Here is a religious sect which hardly had an existence twenty years ago, now numbering its followers by hundreds of thousands. The great majority of these have come out from evangelical churches, and, however indifferent they may have been in their former church relations, they seem now to be possessed by a consuming zeal. Costly houses of worship are erected, and paid for with astonishing ease. Services, which to an outsider seem absolutely wanting in attractiveness, are attended by great numbers of people, and that not once or twice only, but regularly, month after month, and year after year. The strength of Christian Science is greater than can be measured by counting the number of communicants. In many evangelical churches may be found those who, without changing their church relationship, have fully accepted the teachings of Mrs. Eddy. Still others are "tossed to and fro," having gone a considerable distance toward accepting this new faith, but not being, as yet, fully persuaded.

The desire for some explanation of this remarkable growth is increased as we note the enormous inherent difficulties of Christian Science. The tax upon credulity necessary to believe that Joseph Smith told the truth as to the manner in which he received the Books of Mormon, is small compared with that required to accept the dogmas of Mrs. Eddy. To believe that neither the beefsteak which you buy, nor the range upon which it is broiled, nor the platter upon which it rests, nor the mouth into which it is put have any reality, causes a mighty strain upon faith. To many, the fact that Mrs. Eddy is enthroned as a female pope, whose slightest wish is law, and whose dicta are never questioned and from which no appeal is possible, is in itself sufficient reason for rejecting the whole system. Yet, despite the dictatorship of Mrs. Eddy, and with full knowledge of the fact that acceptance of Christian Science theories involves the denial of the constant evidence of their senses as well as what seems to be the plain teaching of the Bible, thousands of honest and fairly intelligent people are yearly added to the already large number of those who look to Mrs. Eddy as an infallible guide.

HALF-TRUTHS HELP TO EXPLAIN GROWTH.

It is much easier to ask for an explanation of this strange stampede than it is to give one that shall be satisfactory. One who can trace and lay bare for inspection all the subtle and eccentric workings of the human mind, who cannot only resolve a belief into its causative influences, but follow the gossamer thread which ties together effect and cause, must be endowed with almost supernatural ability. Some are disposed to say that we have entered a cycle of intellectual disturbance, such as has been illustrated in the rise and spread of Fourierism, Spiritualism, Millerism, or Mormonism, and to give over any further attempt at explanation. This is an easy but not altogether satisfactory way of disposing of the question. Are there not certain facts related to this movement which throw light upon its development? To search for these facts is the purpose of this paper.

At the very outset of our inquiry, let us frankly acknowledge that Christian Science is not a system of unmixed error. Nothing is gained by refusal to credit our opponents with that which is their due. As followers of Jesus Christ we must give assent to some of the things which Christian Science teaches. Patience, unselfishness, purity of heart, brotherliness, love and hopefulness do not cease to be virtues because Mrs. Eddy emphasizes them. When she urges upon her followers the very choices and aims and activities which have formed the themes for Christian sermons and dissertations for eighteen hundred years, we can hardly inveigh against this part of her teaching. However false we may consider the fundamental proposition, "God is all," in the sense in which Mrs. Eddy uses it, still we must confess that it barely escapes being a fair expression of that growing belief in the immanence of God which is so marked a feature of the Christian thinking of to-day. To another fundamental, "God is good," we can only object that it is a weak dilution of the nobler declaration that "God is Love." It is often said concerning Christian Science that "what is true is not new, and what is new is not true." Granted; but that does not dispose of the truth which it contains or rob it of power over human hearts. To say that this is the work of the devil, who is sharp enough to know that unmixed error would not be one-half as useful to him as a compound in which truth is an ingredient, is to substitute assertion for argument, and utterly fails to satisfy candid minds. Explain it as we will, truth is found in this new faith and has its power over the souls of men.

It seems quite probable that at the present time some become Christian Scientists because it is the fashion. It is an age of fads, and this fad is just now the popular one. It is quite the proper thing in some cities to be an avowed believer in the new cult. In Chicago, for example, the congregation which meets in the building recently erected for the First Christian Science Church is not only large but well-dressed. It has in it many men and women of wealth and social standing. That this has its influence upon some is evident from the remark made by a Chicago

Scientist while visiting in a small Illinois village. "I should not care," said she, "to be a Christian Scientist in a small town!" However, this does not explain the early growth of Christian Science, and this motive is probably not consciously present in any large number of cases to-day.

THE FASCINATION OF DOGMATIC STATEMENT.

The satisfaction which multitudes seem to feel in submitting to the authority of Mrs. Eddy, makes it probable that the same cause which takes some people into the Roman Catholic Church accounts in a measure for the growth of Christian Science; viz., a desire to have troublesome questions settled for them. The fact that Mrs. Eddy demands unquestioning acceptance of her interpretations of scripture and unhesitating obedience to her wishes, while repelling some, undoubtedly attracts others. For not a few, independent investigation has no charm. Inveigh against dogmatism as we will, it has ever been and is now one of the most potent factors in the religious world. The average man is impressed with any statement made frequently and with vigor. No inconsiderable number of people in any given community will yield themselves willingly to the guidance of one who claims to know all about everything. Only a small number of people care to work out their own beliefs by difficult intellectual and spiritual processes; it is vastly easier to get them ready made and at second hand. This fact has had not a little to do with the rapid spread of Mrs. Eddy's dogmas.

The fact should not be overlooked that few intelligent and careful students of the Bible have become Christian Scientists. The writer does not personally know of one such instance, neither has he ever heard of one. If it be urged that clergymen are among the number of those who accept Mrs. Eddy's teaching, it may be said in reply that not all ministers are Bible students. It would be well-nigh impossible for one familiar with the principles of interpretation and who has made a patient and prayerful study of the Bible, to accept the fantastic exegesis of Mrs. Eddy. One often hears the declaration from disciples of Mrs. Eddy, "The Bible is a new book to me." No doubt this declaration is true. The Bible of the fathers disappears under the skillful manipulation of Mrs. Eddy, and that which takes its place is most fearfully and wonderfully new. But it is new to many of those who join the ranks of the Christian Scientists because, although Christians, they have not been Bible students. Despite the sermons which they have heard and the Sunday-schools which they have attended, it remains a sad fact that multitudes of church members know comparatively little about their Bibles. They have given little or no time to independent and thorough study of the word of God. We are not trying to locate the blame, but to state a fact. Such people fall easy victims to the biblical prestidigitator. Mrs. Eddy juggles with words, makes an innocent term like "Euphrates" mean "Divine Science encompassing the universe and man," dissects Adam into "a dam, an obstruction," and the credulous spectator believes that this sleight-of-hand performance is a divine revelation. Christian Science would never have become what it is if the members of our churches had been intelligent and thorough students of the Bible.

THE CURES CONVERT.

But Christian Science converts because it cures. All the other factors in its growth put together are insignificant when compared with this. It is safe to say that 90 per cent. of all Christian Scientists are such because they have had personal knowledge of cures wrought by Science healers. A study of "Letters from Those Healed," found in the "Miscellaneous Writings" of Mrs. Eddy, reveals the fact that it was not the self-evident truth of the theories contained in "Science and Health" which brought about the conversion of these people, but relief from physical maladies. One witness states that she bought a copy of "Science and Health," and at first it was like Greek to her; she could understand only a little of it. By persistent reading she was cured of her malady. A gentleman testifies that he had been held in bondage many years by "beliefs of consumption, dyspepsia, neuralgia, piles, tobacco and bad language." He procured a copy of "Science and Health," and, he says, "after some days' reading I was affected with drowsiness followed by vomiting. This lasted several hours; when I fell into a sleep and awoke healed.'' Any one who has attempted to read the text-book of Christian Science will be fully prepared to credit the testimony of this witness as to the drowsiness and nausea. It is doubtful if Mrs. Eddy would have a dozen followers in the entire world were her theories not buttressed by cures wrought.

The fact that Christian Science sometimes fails in attempts to heal, and the other fact that cures are sometimes claimed where none have been wrought, do not prove that all the cases of alleged healing are spurious. People whose honesty is unquestioned, and whose intelligence makes them competent witnesses, declare that they have been healed of serious maladies under Christian Science treatment. As an example of the experiences through which some have passed, the writer submits the testimony of a personal friend, a sincere and intelligent woman. She says: "I had been ill for several weeks, and under the care of a physician. As I was growing worse all of the time, some of my friends urged me to call in a Christian Science healer. I protested vigorously, as I had no faith in the system. The mere thought of employing such an agency was so distasteful that I spent nearly half a day in tears. When I finally yielded to the solicitations of my friends I told the healer who visited me that I had no confidence in her ability to help me; yet after one treatment I arose from my bed, and have not been ill since." This lady is now an ardent Christian Scientist, and her case is duplicated over and over again.

Christian Scientists themselves point to the cures wrought as establishing the validity of their claims. It is not often that one of them will consent to discuss the philosophical propositions of Christian Science, considered in and of themselves. They rest their case not upon argument, but upon so-called "demonstrations." "By their fruits ye shall know them" is the one answer to all objections urged against the theories which they accept. Their reasoning seems to be that if Christian Science cures disease then its theories must be true; Christian Science does cure disease; therefore its theories are true. If we admit the major premise, we must accept their conclusion; for that diseases have yielded to Chrisitan Science treatment in multitudes of instances, is as well established as that quinine is serviceable as a febrifuge.

THE REAL BASIS OF THE CURES.

Having come upon the explanation—in large part at least—it remains to explain the explanation. How are these cures wrought? Mrs. Eddy says that the only true thing which can be said about matter is that it is unreal. It necessarily follows that disease is also unreal. If the body is a "misstatement of mind," the disease can be no more than a state of mind. Eradicate the belief from the mind, and you have effected a cure.

This is mind-cure pure and simple, however much Mrs. Eddy and her followers may protest against this explanation of their work. Every intelligent physician will admit that the mind powerfully affects the body, both in inducing disease and remedially. While there are some things that the mind cannot do, such as giving sight to one born blind, or setting a dislocation, or replacing a limb that has been amputated, it seems quite certain that its range of influence is much wider than has generally been recognized. The mind affects not only the nerves, but the circulatory system and the secretions, and has been known to effect changes in tissue and in structure. Many a man needs only to be told that he looks ill to feel so; while a firm persuasion that we are not sick, or that the remedies employed will certainly prove effective, is of larger therapeutic value than many bottles of patent medicines.

The discussion of the different phases of mind-cure, and of the systems which employ this agency, is beyond the scope of this article. The study of mesmerism, magnetism, clairvoyance, telepathy, mental and metaphysical healing, suggestion and auto-suggestion, hypnosis and autohypnosis belongs to the department of physiological psychology. It is sufficient for our present purpose to call attention to the fact that so far as any of these agencies are employed in healing disease, their potency depends upon securing certain mental conditions. When these mind-conditions have been induced, physical results follow in many—but not all—cases. Christian Science as a therapeutic agent must take its place with other systems which heal disease without the use of drugs.

The assumption that cures establish theories falls to the ground when we consider the number of conflicting theories which would be proved true were this contention well founded. In the city of Chicago is a physician who claims to heal by a process of "vitalization." His avowed theory is that magnetic currents having healing properties flow from his body to the bodies of his patients. He furnishes numerous testimonials from reputable people who swear that they have been healed by him. That he does perform some cures cannot be doubted. Do these cures prove that his theory of "vitalization" and magnetic currents is true? Dr. Dowie, of faith-cure fame, holds that there is a real body and real disease, and that God heals in answer to prayer. Mrs. Eddy declares that neither body nor disease have any real existence. Both perform cures. Are both sets of theories true? That can hardly be, since they are mutually contradictory. Christian Science is only one of many systems which undertake to heal disease without the use of drugs. Its cures are no more real or more wonderful than those wrought at Lourdes or by Schlatter, or by Mental Science. They all have a common method, differing only in details. Thousands of people believe that the Virgin Mary appeared to the maidens in the grotto at Lourdes. They believe this because they have found health through prayer offered in this grotto. Other thousands believe all the absurd and self-contradictory declarations made by Mrs. Eddy, because they have been healed by Christian Science treatment. So great is man's longing for physical health, so keen the gratitude of those who have been physically benefited, that it is doubtful if the time ever comes when men will refuse to accept theories, however fantastic, if presented to them interwoven with healing of bodily maladies.

VI

THE PRECURSORS OF CHRISTIAN SCIENCE

BY FRANKLIN JOHNSON, D.D.

The writers who have preceded me in this series of articles have made it clear that Christian Science is a species of pantheism. All the historic systems of pantheistic philosophy, therefore, are its precursors. But some of these systems are more nearly related to it than others.

TWO SORTS OF PANTHEISM.

It is not easy for any pantheist to regard both God and the universe as simple, sincere and honest. It is the tendency of pantheism in every form to find illusions where the more sober philosophies find substantial reality. The fact that some pantheists have resisted the tendency does not disprove the existence of it, since the great majority of their fellows have been swept away by it. Either matter and motion are the chief realities, and all lofty conceptions of God are illusions, even though innate in the human mind; or God is the only reality, and material things are illusions in a greater or less degree. These are the two extremes between which the pendulum of pantheism ever swings. The chief systems of

75

pantheistic thought which have disturbed the church have tended to the second of these extremes, and Christian Science is now to be added to their number.

Indeed, Mrs. Eddy expressly repudiates pantheism in the first of these two forms, as in the following sentences: "It is pantheistic to suppose that brains are intelligent, or, in other words, that Mind is material. Pantheism is neither Christian nor scientific. The belief that Mind is a product of matter is absurd." "The false doctrine of pantheism that God, or Life, is in or of matter." This materialistic type of pantheism is the only one of which Mrs. Eddy seems to have heard. It is evident from these quotations that when she condemns pantheism she thinks only of the materialistic extreme, and not the idealistic, of which her system of thought, if it may be called a system, is a confused example.

The pantheistic systems with which I have classed Christian Science had their parentage in India, the native home of all hilosophy. Brahmanism, the oldest philosophy in the world —especially as interpreted by the Vedantic school of philosophical and religious teachers—is the earliest precursor of Christian Science. This statement I shall now justify by pointing out several prominent traits of family resemblance.

MATTER AN ILLUSION.

The first is the denial that matter really exists, which, as the reader already knows, is one of the fundamental doctrines of Christian Science. "In the Vedantic school," says Davies, in the introduction to his translation of the "Bhagavad Gita," "all bodily forms or material existences are mere illusion, a temporary appearance, like an image of the moon in water, with which it has pleased the One Sole Being to veil for a time his purely spiritual nature." "The difficult, or rather impossible problem of the origin of matter and of existing forms is set aside by the mere negation of matter."

SIN AN ILLUSION.

A second trait of family resemblance is found in the teaching of the two systems concerning deliverance from imperfection. It can hardly be said that sin, in our sense of the word, is recognized in Brahmanism; but imperfection and wretchedness are recognized, and a remedy for them is provided. As they spring wholly from ignorance, the remedy is knowledge. When a man knows perfectly that all material things are illusions, and that he himself is Brahma, his long journey of successive transmigrations comes to an end, and he enters upon eternal quiescence. Christian Science seems to speak more distinctly of sin, because our age is one in which sin is recognized by millions of earnest persons, and is mourned and hated far more bitterly than are mere imperfection and wretchedness. But Christian Science speaks of sin only to pronounce it an illusion. As Brahmanism knows no real sin, so

Christian Science knows no real sin; and as Brahmanism provides no better remedy than knowledge for human imperfection and wretchedness, so Christian Science provides no better remedy than knowledge for the illusion of sin. We need but to perceive that sin does not exist in order to conquer it. Like disease, it disappears when we deny it a place among the realities of existence.

PAIN AND DEATH ILLUSIONS.

A third trait of family resemblance is found in the denial of the reality of pain and death common to Brahmanism and Christian Science. The author of the "Bhagavad Gita," though he does not hold the doctrine of illusion in its extremest form, yet denies that the body is sufficiently real to enable any one to kill an enemy or to be killed by an enemy. He represents Krishna as exhorting a warrior to enter into battle because he can neither give nor receive any injury. The soul is the man. "He who deems this to be a slayer and he who thinks that it can be slain, are both undiscerning: it slays not, and it is not slain." The "Bhagavad Gita" is a poem, indeed, and some one may object that it is wrong to press its language into the mold of a literal interpretation. On the contrary, while it is a poem, it professes to teach a philosophy. The passage which I have quoted is no mere poetic interpretation of our mortality, like the line in which Longfellow assures us that

"There is no death, what seems such is transition."

It asserts what the author regards as a sober truth, and is employed as the basis of an argument. The body is a mere garment, with no organic relation to the soul; it is put on at what we call birth, and put off at what we call death, but the man is not born, nor does he die. "Wherefore, fight, O son of Bharata." This sounds quite like the declaration of Mrs. Eddy that "there is no birth or death; there is no suffering; it is only the counterfeit of man—the mortal mind—that undergoes these illusions."

THE DEIFICATION OF MAN.

A fourth trait of family resemblance is the identification of the human soul with God, and hence the deification of man. We meet this first in Brahmanism, and last in Christian Science. The devotee of the Vedantic system of religion "must notice the illusory character of all that seems to exist, or all that is besides the absolute spirit, and thereby be in a position to say, I am Brahma, the unchanging, pure, intelligent, free, undecaying, supreme joy, eternal, secondless." The reader has already learned with what boldness Christian Science essays this leap from the creature to the Creator of all.

THE APPLICATION OF THESE VIEWS.

I do not find all the details of Christian Science in Brahmanism, the earliest school of pantheistic thought of which we have any knowledge;

but, as we have seen, its fundamental assumptions are there. The person who accepts these may apply them in various ways. The devotees of Christian Science go further than the most daring of their precursors when they apply them to the healing of disease. The Hindu pantheist, denying the reality of the body, denied, as a logical necessity, the reality of pain and death, but it did not occur to him that the steadfast denial was a means of recovery from the unreal pain and of escape from the unreal death. This last step in the process was reserved for Mrs. Eddy. But the Hindu pantheists made an approach to it. There is a large class of Brahmans who are called Yogis, because they practice Yoga, a combination of ascetic tortures and meditation. One of their sacred exercises is said by them to be "highly beneficial in overcoming all diseases." But it is a mere posture, and the man who assumes it does not attempt to overcome disease by arguing in his mind that there is no disease. Many of the Yogis have gone very far in showing the supremacy of the mind over the body. Monier Williams, in his "Indian Wisdom," describes their achievements as follows: "We read of some who acquire the power of remaining under water for a space of time quite incredible; of others who bury themselves up to the neck in the ground or even below it, leaving only a little hole through which to breathe; of others who keep their fists clenched for years till the nails grow through the back of their hands; of others

who hold one or both arms aloft till they become
immovably fixed in that position and withered to
the bone; of others who roll their bodies for thousands of miles to some place of pilgrimage; of
others who sleep on beds of iron spikes; of others
who chain themselves for life to trees; and of
others, again, who pass their lives, heavily
chained, in iron cages." These and a thousand
other forms of self-torture are practiced with but
little appearance of pain. The denial of the
reality of the body must have encouraged these
ascetic rigors. It is somewhat remarkable that
the Yogis did not take one further step, and try to
find in their theories and usages a means of dissi-
pating disease; though, being acute reasoners, it
is no wonder that they failed to stumble into the
absurdity of Mrs. Eddy, and to teach that cures
are wrought by the steadfast denial that there is
anything to cure.

THE PEDIGREE OF CHRISTIAN SCIENCE.

But, after all, is there anything more than a
series of accidental analogies to connect Hindu
pantheism with Christian Science? Is the one
the precursor of the other historically and
genetically? It is probable that Mrs. Eddy never
heard of Hindu pantheism; yet it is also probable
that it has descended to her by a long and
circuitous route.

Pantheism was a well-known feature of early
Greek speculation. But Greek pantheism tended
to atheism, or at least to the identification of God

with the material universe. If carried out consistently it would make the material universe the reality, and reduce all lofty conceptions of God to illusions. This is the exact reverse of the Hindu tendency. It was not the materialistic Greek pantheism that the early church found most troublesome; it was the Hindu pantheism, imported from India, and mixed with other elements. Nor has the Greek type of pantheism given much trouble to the church in her subsequent history; it has always been the Hindu type with which she has had her chief contention.

Hindu pantheism was brought into contact with early Christianity by the Neo-Platonists. "The philosopher Numenius," writes Möller, "a forerunner of Neo-Platonism proper, sought to gather together the religiously valuable wisdom of different peoples, of the Indian Brahmans, the Jews, the Persian Magi, and the Egyptians." Ever since that time this philosophy of the Brahmans has continued to confront Christianity, for though it has often been subdued, it has never been wholly destroyed, and has been able on several occasions to muster mighty forces and deliver great battles. The last serious combat with it took place as recently as the time of Fichte, Schelling and Hegel. Its chief distinguishing features have always been its assertion of the more or less illusory character of material things, its tendency to overlook or to deny sin, and its deification of man.

Neo-Platonism, borrowing one of its chief

features from Brahmanism, affirmed, says Uhl-
horn, that "matter is the negation of being."
This borrowed doctrine would have led, of itself,
to a shallow doctrine of sin. But, borrowing
another feature from the Persian dualism, Neo-
Platonism affirmed that this "negation of being"
is essentially evil, and the source of the evil which
afflicts the soul, and it arrived thus at its shallow
doctrine of sin by a more circuitous route. Sin
being a thing of the body, the soul can break
away from it by ecstasy or by death. Moreover,
as sin is only a thing of the body, there is the less
reason to hesitate to regard the soul as strictly a
divine essence. When Plotinus was dying he said
calmly to his friends: "I am struggling to liber-
ate the divinity within me." In this manner the
three chief features of Hindu pantheism reap-
peared in Neo-Platonism. Their reappearance in
the German schools of pantheism need not be
illustrated, since those schools are so recent and
so well known.

Though these features, as we have seen, reap-
pear once more in Christian Science, it is not
necessary to suppose that Mrs. Eddy studied any
of the systems from which her theories are borrowed. Indeed, her book is so chaotic, so full of
whimsies, caprices, and impossibilities of thought,
as to render her entire ignorance of all systematic
philosophy apparent. Had her mind been bal-
anced and chastened by reading any treatise on
philosophy, however poor, it would have been
saved from the thousand-fold confusions into

which it has fallen. Nor is it difficult to see how Christian Science has been derived from its precursors by an unconscious process.

New England has long been hospitable to speculations of various kinds. She was unusually hospitable to the transcendentalism of Schelling because it was recommended to her by Emerson and others of her most famous and most beloved sons. Thus for many years before Mrs. Eddy wrote, there was much pantheistic mist and haze floating in the air about her, and she breathed it in her childhood. Later, Boston took up the study of theosophy, and came into contact with a form of Indian thought in which pantheistic elements had a place. These influences, affecting even the newspapers, and through them the whole population, were not unfelt by the founder of Christian Science. Her mind, ingenious though uninstructed, quick to seize a plausible thought, though unable to distinguish the merely plausible from the true, or to connect a series of thoughts in logical sequence, gathered from the atmosphere the few truths and the many errors, crudities, absurdities, and impossibilities which appear in her teachings. A few leading doctrines may be traced through them, and these enable the student of them to recognize them as a species of pantheism. But along with these there is so much of inconsistency, of self-contradiction, and of folly, that the sober reader becomes bewildered, like one who walks in a maze.

VII

THE FUTURE OF CHRISTIAN SCIENCE

BY BENJAMIN A. GREENE, D.D.

It has a future. How far into the years it will project itself, how widely and thoroughly it may spread, no one can predict. This much can be said, and ought to be said, to some who treat the whole question with explosives of disgust: it has a future because it has a vigorous, phenomenally growing present, because, at the heart of this growth, disagree and dissent as we may, there is forcefully evident profound conviction, sincere reverence, such a sense of discovery and victory in daily experience as to make many lives radiant with joy and hope. We judge the future by what the present holds.

ACCELERATING MOMENTUM.

Letters of inquiry sent to persons outside of Christian Science circles, in eighteen representative cities of our country, bring back word that it is still making headway, and in most places with a greatly accelerated momentum. This testimony is confirmed and very largely supplemented by official publications, weekly and monthly, which I have carefully read. In New York State they claim that the number was doubled last year. In 1867 Mrs. Eddy had one pupil. To-day we find in the "Directory of

Christian Science Practitioners" over 2,000 names with city and street address. Every state in the union is found there, also Canada, England, Scotland, France, Germany, and Hawaii and China. The text-book, "Science and Health," has gone through 160 editions, and the call is for more. Three weeks ago Mr. Kimball claimed in his Chicago lecture that Christian Scientists had nearly 2,000,000 instances of healing. Make such allowances as the most skeptical insist upon; even then you have a bulky, vigorous fact striding to-day where it was creeping only yesterday. It has emerged from the silence of contempt. Editors, ministers, medical clubs, legislatures, give it attention. Such an aggressive force is bound to project itself yet further. It has a future.

A FEW PARTICULAR FEATURES.

The uniform report is that their services show a happy and contented company. Something gives them joy. Not many houses of worship have been built; but when they build, they dedicate free of debt. It takes conviction to use the pocketbook in that way. The Sentinel and the Journal give in every issue a large number of letters. Hard-headed business men, as well as women and farmers, open their hearts in testimony; specify as to the healing that has come to their home, sagacity and poise to their business management, and spiritual calm to their minds; and then give name, city, street, number. Just such testimonials are coveted by materia medica

and orthodox churches. They are willing witnesses, and zealous in distributing literature. In
all this we must admit "promise and potency" for
the future.

THE USE OF THE BIBLE.

Though this is eclectic, fragmentary, and, now
and then, flagrantly absurd, still there is a vital
joining upon underlying principles; namely, the
divine immanence, the communion of man with
the Infinite, casting out fear through love, gaining
peace through staying the mind on God, divine
power and promise to heal all manner of
diseases. In the last particular they claim a
fidelity to scripture superior to that of their critics.
They claim to demonstrate the truthfulness of
their position in that they do actually cure disease
without the use of drugs. Leaving the nature of
the cures until later, let it be noted that Christian
Science insists on keeping its teaching blended
with Bible teaching. The mind comes in contact
sympathetically with some of its vitalizing truths.
Such buttressing gives strength that promises
endurance. Even the novel interpretation
awakens interest in the Bible never dreamed of
before. Bringing great truths to bear upon
every-day worries, fears, ailments, and securing
thereby quiet of mind and health, invests that
interpretation, in spite of all imperfections, with a
glory it will be hard to dissipate. Let the man
who applies the whole scripture to his whole self
cast the first stone.

THE LARGE, WAITING CLIENTAGE.

There is no fact more evident than that of the misery and restlessness of mankind through disease. How many million dollars are paid annually for patent medicines to heal "chronic lingerings," and that, too, after other millions have already been paid to physicians! If there be such a thing as "the law of supply and demand," here is a condition of things that will continue to make large demands. And among the people there seems to be a sort of hygienic-messianic expectation. Some are looking for deliverance through a discovery of microbes; others for telepathic channels through which the health of God may surge in upon the ills of man. Christian Science is in search of the man who has tried physicians, baths, travel, climate, and tried in vain. And when healing comes in the last resort, the mind is in a receptive mood for whatever teaching accompanies it. Especially the idea of union with the Infinite, bringing calm, trust, hope, and power over bodily conditions, is a revelation, a joy. Many cling to that, are made staunch converts. The accession of happy converts makes the future sure.

THE FASCINATION OF A SENSE OF POWER.

When homeopathy first came, mothers, not a few, provided themselves with a little book and a box of bottles filled with tinctured pellets. There was a font of healing on the bureau in the bedroom. A kind man would sometimes have an assortment of vials in his pocket, and, as he dropped into the village homes, finding sickness there, would administer relief. The consciousness that he carried with him remedial forces had a fascinating power over his mind, and this, joined with his benevolence, kept him for some time at his philanthropic work. Here it is not pellets in the vest pocket, but thoughts in the mind. "What!" says a man, "can I, by holding my mind in a proper attitude toward the truth, by opening my soul to the incoming infinite love and power, become a channel of health to humanity around me?" "Yes," says Christian Science, and the marvel of that possibility is still in the ascendant.

THE SIDE WHERE CONVERTS MAKE THEIR APPROACH is the one I have thus far dwelt upon. Determination to look facts in the face, facts that are drawing men to-day, and to be candid in the statement of them, is absolutely essential to fair treatment and a satisfactory judgment with regard to the future. To approach Calvinism on the side of Servetus, or Puritanism on the side of Salem witchcraft, would be hardly fair. Nine-tenths of Christian Scientists confess they were drawn to the movement through physical benefit gained.

THERE IS ANOTHER SIDE.

The metaphysical theory as to the unreality of matter and disease and pain is like the bitterness of the physician's pill. Christian Scientists swallow it. They do not care to argue at that point.

As a leading judge of Chicago recently said: "I do not know that we can explain that; but so long as Christian Scientists give to humanity health and hope, you church people ought not to fight them." I most heartily agree with him, if by that term, fight, he means feelingless, pebble-wit flings, poisoned-arrow dartings of speech, vituperative slow-fire, thumb-screw charges of devilish machinations. But I do not agree with him if it means an unquestioning capitulation to all the claims and theories of Mrs. Eddy. Thoughtful discrimination is an alphabetic virtue in listening to any human teacher. Even God himself says, "Come, let us reason together."

THE NEW PSYCHOLOGICAL ATMOSPHERE
which has favored the growth of Christian Science will continue to favor it. A growing conviction as to the power of mind over bodily conditions is "in the air." Christian Science did not create this conviction. Hypnosis, faith-cure, mind-cure, had also been emphasizing it. They all felt the breath of a common evolutionary process. This psychological condition, while fostering the growth of Christian Science, shows it unmistakably to be one of a class (for its cures are certainly duplicated in other schools), and, therefore, discloses

A SWARMING BROOD OF RIVALS.
Mrs. Eddy claims the Bible as her only source of knowledge, and yet frankly confesses that her "favorite studies were natural philosophy, logic and moral science." The anthem of Emerson's

"Over-Soul" had been reverberating through
New England for twenty-five years. The very
year she began, 1866, was the year Dr. P. P.
Quimby died, the man whose original contribu-
tions in this line form the basis of such work as
Horatio W. Dresser gives us in "The Power of
Silence." Indeed, from time immemorial, the
marvelous power of mind over body has been
acknowledged. "A merry heart doeth good like
a medicine." "It is part of the cure to wish to
be cured." " 'Tis the mind that makes the body
rich." Surely Solomon, Seneca, Shakespeare had
inklings. Physicians who use drugs, include, in
proportion to their good sense, mental conditions
in diagnosis and treatment. Psychologists have
found, in hypnotism, an open door into this whole
wide realm. Common humanity has made it common talk for ages, "If you charge a man's mind
with good news and good cheer, you aid digestion,
quicken circulation, begin to build physique over
again." This thought has had currency; but it
was in nickels and dimes. What has long been
needed is larger pieces; halves and "cart-wheels,"
$5 gold pieces and bank notes with two or three
figures on them.

THE CHRISTIAN CHURCHES OF THE FUTURE.

Thoughts are vital forces. Truth is thought-
force bringing freedom. Words are "storage
batteries." "The Kingdom of God is within
you," and connected by spiritual telephone with
the King eternal. In his presence is fullness of

life. These are the quickening revelations now whispering their way into larger and larger acceptance. In such literature as "Ideal Suggestion," "In Tune with the Infinite," "The White Cross Library," we have a new pharmacopœia widening out into the future. Christian Science will undoubtedly share largely in this expansion. It must be given credit for daring to thrust the larger pieces of currency into the broadest circulation thus far.

Christian churches will be made to feel this call to give the power of God a broadening application to our incarnate life. What psychologists are saying with united emphasis, what mental healers are now actually proving, will come with cumulative urgency, and demand that we bring out a class of scripture long time held in abeyance, and give it current hygienic value. "The prayer of faith shall save the sick." God shall quicken our moital bodies, while they are alive, through the divine tonic of his Spirit's indwelling. Not that disease and death will be banished, not that physicians and surgeons will become anachronisms, but the ills of flesh may be forced into diminished area and the glory of the Christian's final victory be preceded by anticipatory thrills of triumph. Daily living will be more of a victory and less of a defeat. When this comes true, Christian Science will find one of its largest tributaries cut off, and the church of Christ, more potent by reason of added and emphasized truth, will have become the better prepared for new triumphs.

DISINTEGRATING INFLUENCES ALREADY AT WORK.

Growing up around the "Mother Church" in Boston are schools of healing which retain the impersonal elements and ignore Mrs. Eddy. All through the country, in larger centers, this is the case. In Chicago "The Exodus Club" has for its chief lecturer Mrs. Gesterfeld, who was formerly Mrs. Eddy's "star-pupil," so President Trude told me. Jane W. Yarnall writes a book, taking the same ground. In Washington, many persons retain church membership and practice what they term "applied Christianity." These are only straws in the wind. From words of complaint, scattered through Christian Science literature, charging plagiarism and ingratitude, and from the more rigid restrictions put upon the public meetings, it is evident that the disintegrating process is already seriously felt, and what it threatens is more seriously feared. The Come-outers say that

MRS. EDDY CLAIMS TOO MUCH FOR HERSELF PERSONALLY.

While acknowledging her as a discoverer, they think that truth, "broad as the cosmos," cannot "be so compressed as to flow only through one channel." Christian Science is "autocratic, rather than democratic"; "polity and ritual in every detail are shaped and directed by a single will." And the spirit of all this an "outsider" easily discovers in reading Mrs. Eddy's works, e. g. "There is only one Christian Science, as there is only one Truth." "The wise Christian

Scientist will commend students and patients to
the teachings of 'Science and Health' and the
healing efficacy thereof, rather than try to center
interest on himself." "*The Bible and my books
mislead no one.*" The sermon on Sunday is
made up entirely of extracts from her writings
and quotations from scripture. There are no
lecturers recognized except those appointed and
absolutely governed from headquarters. She
out-popes the pope. Recently she ordered Dr.
Geo. Tompkins from the lecture platform to the
more private sphere of healing, and he meekly
states his acquiescence in public print. A
humbler instance of submission came out in the
Christian Science Weekly last November, in connection with the question whether clippings could
be read in the Wednesday evening meeting. A
"reader" had severely chided a person for doing
that. If it be contrary to Mrs. Eddy's wish, this
person wrote to the editor, he would "beg her
pardon and promise never to do it again." When
Mrs. Eddy dies, who will inherit her claims to
infallibility? Will not that event precipitate the
process of disintegration?

THE "UNREALITY" HANDICAP.

1. As to matter. Idealism may do for theoret-
ical discussion, but take it out into work-a-day life,
it will not work. Mrs. Eddy insists, for the mil-
lionth time, that "matter is unreal"; "knowledge
gained through the material senses is *only
illusion.*" Stab a man; red blood flows; death

robs a family of support. Illusion! People may swallow that absurdity because of physical benefit gained; but, in the long run, common sense will resume operation. It is sure to resume when the practical Yankee instinct recovers and gets back of the perception which enables them to see that Mrs. Eddy makes the very shrewdest use of the material printing-press for the advocacy of her unreal "unreality," and counts on good eyeball vision to take the message in. Jupiter sometimes nodded. And even now some are sorry to see that their ideal leader allowed herself to be caught nodding in that she sanctions the embossing of her features on a silver spoon and urges each one of the 300,000 disciples to purchase at least one material souvenir at $4 apiece, and where they can, to take a dozen for the use of visiting friends. The "unreality" theory is not essential to the cures. Those who reject or abandon that phase of "The Science of Being," and yet hold to principles common to several schools, still succeed in mental therapeutics. Even "Father John," of the Russian Church, is to-day healing with extraordinary success. People run out to meet him as they ran to meet Paul at Ephesus. He is the sainted miracle-worker of Kronstadt. This tying up of healing power with the "unreality" theory reminds me of a woman I once knew who thoroughly searched the Bible with a view to successful defense of the doctrine of the sleep of the dead. Because she had spiritual quickening through contact with much Bible truth in prolonged study, she attributed it to that particular doctrine, and hence that doctrine must be true.

2. As to sin. In "Unity of Good" Mrs. Eddy speaks of the "*illusion* which calls sin real, and *man a sinner needing a Savior.*" She began with the unreality of disease. She would drive the thought of disease out of mind, insist on letting the health-thought come in, fill her, and deny the intruder right of place. In stress of battle, there at that victorious point, she thought she had a revelation of a wide-reaching principle. "God is All; God is Good;" therefore, "there is no such thing as disease or evil," both are "illusions." Danger here for the future. Conditional sentences may be thrown in freely; positive injunctions to be kind, just and pure may be added; but that underlying principle of the "unreality of evil" plunges a dagger through the Bible doctrine of a man's individual accountability and lets out the very heart-blood of Christ's distinctive teaching. To see what might be the worked-out results of that teaching in remote time and place, we have only to turn to history and read of Antinomian license. The opposite extreme, holding true now, as I have been told, in isolated instances, goes to show just that weakness of extremes which warns of a possible pendulum swing later.

HOW SHALL WE TREAT CHRISTIAN SCIENCE IN THE FUTURE?

Treat it kindly as you would a brother. Face all the facts. Frankly acknowledge every iota of good. Just as frankly call attention to what you sincerely believe to be error and danger. To ignore it, boycott it, set yourself dogmatically against it, saying it is nothing but a tissue of deception, is as unreasonable as you try to make out Christian Science to be. The ordinary church member with his worry and fear, with a religion that does not give him a baptism of strength and joy to-day, with lips unable or unwilling to witness for the divine power, may learn a very useful lesson from Christian Scientists. "Tell how great things the Lord hath done for you." Mrs. Eddy may claim, as she does, that she "introduced the first purely metaphysical healing since the apostles' day," that scripture itself "gave no scientific basis for demonstrating spiritual principle until the Heavenly Father saw fit in 'Science and Health' to unlock the mystery of godliness." But she is mistaken. She is human, and she shows her frailty as did Luther when he said that the Epistle of James was an epistle of straw, as did Miller when he fixed upon 1843 for the second coming of our Lord. Gautama was human. Mohammed was very human. Both of these systems have lasted through centuries. Mrs. Eddy says: "Centuries will elapse before the topics of 'Science and Health' are sufficiently understood to be fully demonstrated." Her name may ascend and find place amid the century constellations. But I have my doubts.

VIII

THE ERRORS OF CHRISTIAN SCIENCE

BY CEPHAS B. CRANE, D.D.

BY CEPHAS B. CRANE, D.D.

Few of the readers of The Standard could be more interested than myself in the foregoing able articles upon Christian Science. I was living in Boston, which was then her city, when Mrs. Eddy began to attract public attention to herself and to her system. At a later period we both had our homes in Concord, N. H., where I often saw her in her daily drives, and where I became somewhat acquainted with the story of her life and achievements.

It is plain that the writers of the articles sought to be fair, even generous, in their estimate of Christian Science. They showed the chivalric spirit which ought always to characterize men when they pass judgment upon the achievements of women. I cannot see how Mrs. Eddy, or any of her followers, can bring a charge of injustice or harshness. All that is true and good in Christian Science was warmly commended. Probably I am not the only reader of the articles who perceived, or imagined, that the writers did not at all points agree in the interpretation of the system. But this is not at all to their discredit.

98

The fault is with the system. It is not self-consistent, and it is nebulous. The scientific and philosophical and theological language of it is to an amazing degree indefinite and elusive. There is constant contradiction in terms. Familiar words are charged with unfamiliar meanings. The thoughtful and well-instructed reader is perplexed that he cannot get out of the fog into clear air. He is baffled at every turn. Emerson is sometimes called obscure, for the reason that he leaves the reader to supply the minor connective sentences. But no most accomplished reader can supply the minor connective sentences of the text-book of Christian Science. There is a perpetual to and fro movement between what appears to be logic and what is sentimentalism and mysticism. If it were either logical or rhapsodical or apocalyptic, one might get at the gist of it. But the trouble is that it is a compound of all that these adjectives signify. Formidable postulates are verified by fancy and emotion. It is vain for one who is balked by an apparent conclusion to go back to discover the premises. The premises are not there. The conclusion is an affirmation without visible support. So one reads on, every moment more hopelessly bewildered.

The palpable truth in Christian Science is that what the uninitiated multitude call the body is to less or greater degree affected by the mind. Many pains and aches vanish at once when we are made to believe that there is no cause nor reason

for them. I have more than once had such
deliverances when my doctor has told me that
nothing ailed me. Indeed, only last week I
delivered a man in precisely this way. Hysterics
and hiccough can be abolished by an imperative.
It is perfectly true that certain states of the body
are determined by certain states of the mind.
But this is no new discovery. The doctors have
helped patients wonderfully by bread-pills, and a
few drops of water disguised under the cloak
of a harmless essence. If they say bread-pill and
pure water, their prescription fails; but if they
say *panis pilula* and *aqua pura*, the prescription
works like magic. As to the more obstinate evils
of ulcer and cancer and a broken leg—but that is
another story.

The exhortation to trust and serenity, constantly given by the hierophants of Christian
Science, and usually obeyed by the neophytes, is
accounted a praiseworthy exhortation. But trust
and serenity, if they rest upon a false foundation,
are pathetic. "I am perfectly well," said a
young woman to me at our seaside hotel, who
was constantly reading the text-book of Christian
Science, and who kept her neighbors awake in the
night with her distressing cough. Poor girl!
She returned to her home in California to die.
Trust and serenity are really admirable only when
they are warranted by the facts in the case.
Otherwise, they awaken in us an unspeakable pity.

Christian Science claims to be a science, a
philosophy and a theology. In my judgment it is

no one of them. As a science it begins by making all other sciences, particularly the physical sciences, preposterous. For according to it, since matter is an illusion, and all is mind, astronomy, geology, botany, paleontology, medicine, are baseless. The men of eminence in these sciences waste their time upon "chimeras ruminating in a vacuum." Nor can the Christian Scientist save himself here by classing himself with accredited spiritual monists. For the clear-eyed monist, even though he affirms mind to be the one and only substance, does not dream of saying that the ordered phase of being to which we give the name of matter is an illusion. He calls it a reality. There may be two spiritual systems, the one as real and permanent as the other. The triangle is an idea, and the circle is an idea, but neither can abolish the other.

Meanwhile, how can there be mental science when personality, even the personality of God, vanishes into the impersonality of a principle? Or how can there be mental science when the axiomatic first truths upon which all science must build are flatly denied?

Christian Science is not a philosophy. It is subjective idealism gone mad. Professor Bowne, of Boston University, and Professor Royce, of Harvard University, are both eminent exponents of philosophical idealism. They construe what we call matter in terms of mind. But they would never dream of calling that system of things, that established and permanent order, to which we give the name of matter, an illusion. They stoutly affirm its reality. They bid us respect it, and conform our action to it. They eat and drink, knock out the stones that get wedged in the shoes of their horses, send for the doctor and take his medicine, without a spasm of uneasy feeling that they are contradicting their own philosophy. But when the Christian Scientist does these things he practically abjures his philosophy. Why eat and drink if you have no body with a hungry and thirsty stomach? Why knock out the stone under your horse's foot when there is no stone and you have no horse? Why take medicine, when disease is an illusion? True, the Christian Scientist refuses the medicine, even if he has to die for it. But why does he not also refuse the butcher's meat and the bread? Why does he not allow his foolish horse to go on limping under the delusion that he has a stone under his hoof? In theory he denies the existence of what goes by the name of the material world; in practice he respects it as fully as do all the rest of us. Of what value is that philosophy to which the conduct of life cannot possibly conform?

Christian Science is not a theology. How can it be a theology, when God is in one sentence falteringly named a person, and in the next sentence stoutly affirmed to be a principle? Person or principle—which? Speak plainly, my friend, that we may know "where we are at." It is clear that here our friend is floundering in the slough of a misunderstood pantheism.

The theology of Christian Science, in its effort to be true to its idealistic philosophy, has the temerity to pronounce sin an illusion. That surely is an astounding assumption. But philosophical idealism does not require that one thus fly in the face of things. Professors Bowne and Royce blister sin with fierce invectives. They deny that it is an illusion. They affirm its reality, its punishableness, man's need of deliverance from it.

On the whole, I should say that Christian Science ought to go to school. Let it take a thorough course, under competent teachers, in physical science, philosophy, and theology, and it will slough its absurdities, become conformable to reason and faith, and become capable of being fairly understood by the average man.

IX

SOME INCIDENTS OF PRACTICAL CHRISTIAN SCIENCE

BY E. S. PLIMPTON

In the previous articles a most generous attitude has been maintained towards Christian Science. Curative results have freely been granted, and the successes claimed have not been disputed. In fact, one of the greatest reasons for its spread has been given: "It cures!" Indeed, this is cited as the cause. But in all fairness may not another reason be stated? It pays. '

IT IS PROFITABLE.

While hesitating to make the charge that selfish considerations are consciously paramount, is it possible that any system shall be exceedingly remunerative to its promoters and they be wholly uninfluenced by this profit? I have before me the Christian Science Sentinel, the great organ of the system, published in Boston, dated February 16, 1899. In this paper nine columns are devoted to the recital of the history of the litigation, in which Mrs. Eddy is the complainant, of the violation of her copyright of the book "Science and Health," and of other of her writings, and a copy of the court decree granting her a perpetual injunction in the case is given. The copyright is thus sustained, and she is sustained in her claim of exclusive right of publication and sale and the consequent profit therefrom.

That this is no valueless decree is shown by the accompanying price list of her works, among which "Science and Health" is quoted at $2.75 per copy for twelve or more copies to one address, or $3.18 singly. This is for the cloth edition. Since books of equal volume, style of binding and paper are freely on sale at 75 cents, there is left a matter of $2 or so to the credit of the copyright. In the publication of 160,000 copies at this rate the profit is clearly enormous.

But this is not the sole source of income, for other publications are sold in lesser number but with corresponding margin of profit. Also, to all this must be added the substantial fees charged for teaching. These are all Mrs. Eddy's own personal property.

The class-teaching may have ceased, but the books are a veritable mine of wealth. No wonder she protects them even by law, and no copy can ever reach an ignorant world except through the channel of her copyright. But this "Science and Health" is accepted and used as a companion to the Bible. Think of a royalty on every copy of the Word of God! Think, too, of the vast difference in price. A full, complete copy of the one may be had for 25 cents; of the other with far less volume the cost is $2.75. Most surely were the

air that we breathe and by God's free bounty
have in abundant supply, subject to such man or
woman's control, every inhalation would be
taxed.

IS A SOURCE OF PROFIT TO OTHERS.

That the teaching of Christian Science as well
as its practice is profitable is seen by this illustra-
tion, which is only one of multitudes of instances,
and I will here state that no incident but those
personally known to the writer will be cited.

A merchant gave up a paying business to
become a teacher of Christian Science. This
was some years ago when teaching was popular.
He went to Boston and took a course of lessons
from Mrs. Eddy herself, at the regular price of
$300 for twelve lessons. He then returned home,
secured classes numbering in some cases as high
as twenty, at $50 for the twelve-lesson course.
Thus six pupils pay his own tuition, and the
remainder of receipts is largely profit. In their
turn these pupils become, or seek to become,
teachers and practitioners, the former at $5 per
pupil, and the latter at $1 to $2 a treatment. The
money profit is great, if at all successful, and
both this success and profit are contingent upon
the spread of the desire for instruction or treatment. Thus, a strong motive exists for diligent
promotion of the doctrine.

Has this motive been without effect? The
writer's experience has been that those practitioners of the system who hold to it most firmly are

of the class of persons that we recognize as being thrifty, able to make the most of what comes in their way. Again I ask, is the fact of profit without its influence? Is not the world full of examples of the promotion of that which promises returns of great profit to the promoters? May not, indeed, the question of profit compete with, if not overshadow, other incentives? In what feature of this phase of the question is there harmony with the command of Jesus: "Freely ye have received, freely give"?

But it may be said, "We give adequate returns for our teaching and our treatment. People need not come to us; we do not compel them." But is not the promise of healing to one in suffering in some degree a compulsion? I know of no class of practitioners who are so profuse in such promises. Moreover, there is always a hazy mystery maintained that of itself is exceedingly fascinating. It partakes of the mystery that describes the drugs of a prescription in an unknown tongue, or that adds to the awe of a religious service by reciting it in Latin. Clothed in verbiage and vaguely worded ambiguity, its power is great.

FAILURE FREQUENT.

But these promises are not always fulfilled. I could mention instance after instance within my own knowledge. One or two of these may be given. A woman with a diseased finger that persistently resisted medical effort was told that it must be amputated. The surgeon came from a distance to operate, but a Christian Science practitioner, on hearing of his coming, went to the home of the poor woman and so worked upon her as to gain her refusal to submit, and the surgeon went home without doing the work. The Christian Science woman assured the patient that she could and would restore the hand and save her finger, and at once began treatment that went on for some weeks, the disease in the meantime constantly spreading until at last a greater part of the hand was involved, and the surgeon was again sent for and left her but the small stub of a hand, which then healed and saved her the loss of her arm. This poor remnant of a hand I have more than once seen.

Again, the two most prominent women who teach and practice Christian Science in this community are both widows within a year from this date. But the strangest part of all is the complacency with which a death under Christian Science treatment is viewed by those in charge. Surprise, grief, regret seem utterly absent, as possibly they may well be, for death is unreal, and there is no cause for grief, for he has only "passed on."

STRIKES AT COMMON SENSE.

A young lady, a teacher in our city schools, had a boil upon her fair cheek. As it inflamed and grew, it attracted much notice, and the more so because it so disfigured her beauty. She, when

spoken to, declared it was not a boil; it was but
the "belief of a boil," therefore she allowed it to
take its course unmolested except by the "negation" that she and her friends brought to bear
upon it. But the boil flourished, attested her
sense of its presence by the tears which the pain,
in spite of her, would cause, it being so near her
eye, suppurated, scabbed, and dried up all
untouched, and left a purple spot as a protest
against neglect.

WEAKENS VERACITY.

A merchant whose word had never been questioned, bargained to exchange with another certain real estate. The other party offered to make
a payment at the time which should bind the
bargain, but was assured that his word was good,
and that each could trust the other to fulfill his
part. After some days it was known that the
merchant had sold his land to still another man.
At once the first party went to him, and having
learned the truth, remonstrated. "But," said the
merchant, "you can't blame me, I got $200
more." "But you gave me your word."
"Well," impatiently, "there isn't anything that's
real." This man was and is a Christian Science
man. What becomes of business integrity with
such views of obligation?

WOULD DESTROY SCIENTIFIC SKILL AND KNOWLEDGE.

A daughter was taken from our city school
because physiology and hygiene are in the course
of study. For months she had no instruction, but

later was sent where, by payment of tuition, such study could be avoided. Such a course is the legitimate fruit of the denial of disease or disability, and the consequent rejection and disuse of surgical or medical skill. One cannot fail to wonder what course of treatment would be adopted in case of a cinder in the eye. Would it be left to stay or come out, as the case might be? Or, if it were too deep ever to come out or to be gotten without instruments adapted to the purpose, would they be used?

A poor drunken young man, whose parents lived on one of my farms, had a cancer on his cheek. It was malignant and corrupt. On a hot summer day he lay in a drunken stupor and the unprotected sore was covered with flies. Maggots, too, were rolling and tumbling over each cther in plain sight and also deep down in the flesh. If this young man in his desperate condition were placed under Mrs. Eddy's treatment, what would she do? Nay, if he were her own son, and there are many as wayward sons of as noble mothers, what would she do? Would the harmless, at least to the feasting vermin, negation or demonstration satisfy her mother-heart? Would it any mother's heart?

Are microbes that can be seen only with the aid of a glass less real than those seen by the eye in this man's cheek? If actual remedies or applications are used to destroy or counteract the one, should they be discarded in the other? Many diseases are germ diseases. But, must Christian

Science ignore those other diseases whose cause is vice and crime? Are the elite of diseases to be reserved as its special own?

Will a demonstration tie a severed artery? Will it suffice in place of the necessary severance and closing of the artery at birth? But, if not, then surgery is admitted, and once permitted to compete for success, who has any doubt as to its outcome? In a railroad accident where there were the maimed, the fractured, the injured externally and internally, would a detachment of Christian Science healers receive a cordial welcome from the wounded and suffering if such were sent on a relief train instead of surgeons and actual doctors to their rescue? Why do not Christian Science practitioners establish their claims for relative efficiency as do other schools of healing, when they, in large city hospitals, take their turns of the cases of accident and disease, and a careful and correct record is kept of every case from the beginning to the end of it?

A FRUITLESS STRUGGLE.

An infant was sick; a Christian Science practitioner, a neighbor, called and offered to treat the child. It was evening, and the child lay in its cradle in a darkened room. The treatment began. That cold, repellent look that indicates the mental fight with an unseen foe, was on her face. After a long time the father in his grief felt that he must see his child. He went to the cradle and was horrified to find his darling cold and almost stiffened in death. During nearly the whole time of treatment the dear child had been dead. The forbidding aspect had kept death at bay not an instant, and the silent operator had been treating a corpse.

THE REAL DANGER.

But the most pitiable of all, because of its far-reaching consequences, is yet to be told. Eternal salvation is at stake. Without faith in an atoning Savior the gospel is an unmeaning word. The atonement of Christ is of no avail to those who willfully reject it or knowingly and of purpose deny it. But who can affirm or believe in that which he declares never took place, and for which there is not only no necessity but no possibility? If there can be any surer way to keep a soul away from its only Savior than to deny that there is a Savior, and to assert that by no sort of means could such a Savior be needful or even possible, such way has not been devised. A soul without a Savior is a lost soul.

I cannot believe that Mrs. Eddy, or other Christian Science teachers, realize the consequences of the rejection of the Lord Jesus Christ, as the atoning Savior, or the failure to accept him if once made known. "He is the propitiation for our sins." So to interpret scripture as to do away with salvation through faith in Christ is to pervert, to destroy that blessed way of life, than which there is no other. An ostentatious

use of the Bible can make no amends for destroy-
ing its very essence.   That soul which feels a
confident security for eternity because of the possession of a certain undefined benevolence of
feeling toward God, which he persuades himself
God feels toward him, is in great danger.
Nowhere in his word can we find any approach to
the Holy One by sinful man except through a
Mediator.   To do away with Jesus Christ as the
atoning Mediator is to bar the only way of access
to God, the Eternal.   Whatever does this is fatal,
be it fashionable, attractive, or clothed in the
form and character of the great adversary of
human souls.   It is just as deadly to go over
Niagara in a gilded boat as upon a raft of logs.

X

THE FORM AND SUBSTANCE OF CHRISTIAN SCIENCE

BY O. P. GIFFORD, D. D.

Christian Science has a hard time getting itself understood; it has a wide hearing, but finds few with understanding hearts. It stands, like the temple, with its outer court of the Gentiles, court of the women, court of Israel, and its holy place. The outer courts are crowded, but there be few who enter the holy place, and stand with the high priestess face to face with the truth. There are many who accept and profit by the health, few who understand the science of the movement. Many catch glimpses of the truth that lies like the blue sky beyond the clouds, but few have reached the heights and live above the clouds always in the light. This difficulty of understanding lies first in the form; second, in the substance of Christian Science.

DIFFICULTIES IN THE FORM OF CHRISTIAN SCIENCE.

Its form is the English language, but the words are used in a foreign sense. M. Taine says of "Paradise Lost": "This Adam entered Paradise via England. There he learned respectability,

and there he studied moral speechifying. Adam
is your true pater familias, with a vote, an M. P.,
an old Oxford man, consulted at need by his wife,
dealing out to her with prudent measure the
scientific explanations which she requires." So
we may say of "Science and Health." It reached
Boston via the Tower of Babel; it confounds a
fairly sensible tongue and thus confuses thought.
The glossary, added to the classic of "Science and
Health," attempts to "elucidate the meaning of
the inspired writer." "It contains the *metaphysical* interpretation of Bible terms, giving
their *spiritual* sense, which is also their *original*
meaning." It would be difficult to find in
another sentence metaphysical, spiritual and original serving as synonyms. Such use of words
reminds of Hamlet and Polonius:

Ham.—"Do you see yonder cloud that's almost
in shape of a camel?"

Pol.—"By the mass, and 'tis like a camel,
indeed."

Ham.—"Methinks it is like a weasel."

Pol.—"It is backed like a weasel."

Ham.—"Or like a whale."

Pol.—"Very like a whale."

This shifting, cloud-like use of words is very
puzzling; it changes the frontiers of thought so
frequently that one scarce dares invest lest his
property slip from his grasp. Adam is defined in
this metaphysical, spiritual, original glossary to
mean:

"Error; a falsity: the belief in original sin,

sickness and death; evil; the opposite of Good, or
God, and his creation; a curse: a belief in intelligent matter, finiteness and mortality; dust to
dust; red sandstone; nothingness; the first god
of mythology; not God's man, who represents the
one God, and is his own image and likeness; the
opposite of Spirit and its creations: that which is
not the image and likeness of Good, but a material
belief, opposed to the one Mind, or Spirit;
so-called finite mind, producing other minds, thus
making 'gods many and lords many'; a product
of nothing, as the opposite of something, an
unreality as opposed to the great reality of
spiritual existence and creation; a so-called man,
whose origin, substance and mind are supposed
to be the opposite of God or Spirit; an inverted
image of Spirit; the image and likeness of God's
opposites, namely matter, sin, sickness and death;
the antipodes of Truth, termed error, the counterfeit of Life, which ultimates in death; the
opposition of Love, called hate; the antipodes of
Spirit's creation, called self-creative matter;
Immortality's opposite, mortality, that of which
wisdom saith, 'Thou shalt surely die.' This name
represents the false supposition that life is not
eternal, but has beginning and end; that the
infinite enters the finite; intelligence passes into
non-intelligence, and soul dwells in material
sense; that immortal mind results in matter, and
matter in mortal mind; that the one God and
Creator entered what he created, and then disappeared in the atheism of matter."

Such a mosaic of words without a pattern of thought bewilders. The ancients traced figures of men and animals in the starry heavens, but imagination fails when we try to arrange these words as thus related to each other so as to form a pattern of thought. With Polonius we are puzzled; have we a whale, a camel or a weasel?

> "Know then thyself, presume not God to scan:
> The proper study of mankind is man."

But such a man defies study, and yet this man is the one most of us know most about by experience.

Elias is defined to be Christian Science. The Holy Ghost is Divine Science. The New Jerusalem is Divine Science. Thus we have an historic character, a prophetic city and the Holy Ghost all meaning the same, and that same Christian Science is the modern metempsychosis, it changes form so often through its strange use of words that a man questions if he be not his own ancestor, and may yet be his own descendant.

DIFFICULTIES IN THE SUBSTANCE OF CHRISTIAN SCIENCE.

Puzzling as the form is, the substance is still more baffling. It is neither gnosticism nor idealism, nor pantheism, though bearing a family resemblance to each in turn. It is a revelation. "Christian Science reveals incontrovertibly that Mind is All-in-all, that the only realities are the divine mind and idea." . . . "God, who at sundry times and in divers manners spake unto the fathers through the prophets, hath in these last days spoken unto us by his Son." Whatever truths there may have been in the broken, fragmentary revelations of the past were all gathered up in the fuller revelations of the Son. But this revelation of and through Christ was also incomplete. He promised another Comforter who should lead into all truth. Christian Science is this promised Comforter. As Christ's revelation interpreted all previous revelation, so Christian Science interprets Christ and previous prophets. This is "a final revelation of the absolute principle of Scientific Mind-healing." This final revelation is of God and man and their mutual relations. This revelation is that "erring, mortal, misnamed mind produces all the organism and action of the mortal body," and that "all real Being is the divine Mind and idea; that Life, Truth, and Love are all-powerful and ever-present; that the opposites of Truth—called error, sin, sickness, disease and death—are the false testimony of false material sense; that this false sense evolves, in belief, a subjective state of mortal mind, which this same mind calls matter, thereby shutting out the true sense of Spirit."

A revelation is to be received, not reasoned about; welcomed, not debated; lived, not thought. This revelation gives God's point of view. Mrs. Eddy stands at the center of the universe and looks out. Thus standing, there is no time, no succession in events, no space-relation. When she says, "There is no matter, no sin, no sickness, no death," she states what is not to God. Now and then a prophet caught a glimpse of God's purpose and foretold to men what was present to God. Christ put God's present purpose and thought into the future tense for man; Paul saw a redeemed universe as it was in God's thought, but always spoke of it as in the future for man; but Mrs. Eddy, standing by God's side, deals in the eternal present. To him the verb of life has but one tense, the present tense. She does not write God's thoughts in man's grammar, but seeks to compel us to share God's thought.

A PARAPHRASE.

Picture a man standing before a large mirror. To him there is nothing but himself and his image or reflection. It has no meaning aside from him; its business is to reflect him. As long as he stands thus before the mirror in the light, its motions correspond to his motions. Suppose this an eternal mirror, faced by an eternal man, casting an eternal reflection, then this man and his reflection are all there are in the universe to him. Suppose this reflection endowed with power of thought, of choice, of will, suppose it to come to self-consciousness, to begin to think it was something in and of itself. It begins to think its own thoughts, to make a world for itself. The man thus standing sees simply his own reflection; to him this change in the image is not real; he sees himself in the mirror, but the reflection sees itself and loses him. Whatever comes into the life of this image is not real to the man, and so not really real. The world of the image is but a dream; his thoughts are not real thoughts, for the only reality is the man and the reflection which he sees.

God is the eternal being, truth, life, principle. He is conscious of himself and his idea, his reflection. This idea is as eternal as he is. It always was, always will be; it has no real being except as it reflects him. God is Spirit. Man is God's spiritual reflection. This spiritual reflection came to self-consciousness, lost God-consciousness; this loss was his, not God's. To God he is still the spiritual reflection, God sees him as he always has seen him. This passing phase of self-consciousness is unknown to God, hence unreal in any right sense. Coming to self-consciousness, thinking himself to be a real thing apart from God whose reflection he is, this idea of God's began to think its own thoughts. Himself a thought of God, instead of reflecting God, he began to express himself. God's reflection is spiritual, the real man; man's reflection is what we call matter. This reflection of man's is real only to himself, not real to God, so not true. The spirit of man is God's thought, idea, reflection; the body of man is man's idea, thought, reflection; it has no reality to God, hence no reality in the right sense of the word. Having thought a body, man goes on to think a world to fit the body, a universe as a part of the world. All that lies within the horizon of the sense-life is the reflection of mortal mind, and is unreal to God. A building is the expression of mortal mind; it is in thought before it is in sight. The bricks and hewn stone and shaped timber are also the expressions of human thought. Bricks, cut stone, hewn timber are first thought out; but the clay, the stone in the quarry, the trees on the hillside, are also expressions of human thought. God thought and thinks man's spirit, man did think God when he was true, now he thinks matter, in many forms, and is false. He has gone into the far country, is wasting his substance in riotous living, is thinking in terms of wine and swine. When he comes to himself he will come to his father, then he will begin to live again. This life in the far country is unreal, is death. God sees only the son of his love, sees him returning, sandaled, ringed, clothed, kissed, restored. To the sculptor the only reality is the form he sees in the stone; the stone is only the accident, the idea is the substance. To the artist the idea in his mind is all, he does not dwell on the canvas, the color, but sees all the time the perfect picture. To the real thinker, language is not thought of, the thought is all. Thus God deals directly with the spirit of man; to him matter is not, it never was the expression of his thought. Thus it follows, that whatever is thought or suffered in matter is unreal to God. Pain, disease, sickness, death, are only human experiences in a realm that has no existence to

God, so they have no reality to God. These experiences are real to man, when man is false to God, but unreal to God because he deals only with the true, the real. There is no sin to God; because sin is not in God's thought, only in man's thought when he is false to God. Man as God thinks him does not, cannot, sin; whatever man thinks when he is false to God is not real to God. So far as man reflects God he is real to God; when man thinks away from God, his thoughts, in terms of matter or of sin, are unreal. The secret of health is in thinking God, reflecting God; God's life thus possesses man's spirit, matter loses its reality, sin, sickness and death disappear from experience, blotted out as the clouds are by the sun, or evil by good.

A lake reflects the sun, is ablaze with its glory, lives in its light. Presently a mist rises from the lake and shuts the sun out, the mist is more real to the lake than the sun is. Instead of expressing the sun, the lake now expresses itself, its own expression shuts it in; shore, sky and sun are all blotted out. The lake knows only itself and its expression, its mist; the sun shines right on, the mist disappears. The man God made reflects God, man as we know him is caught in the fog of his own making, in matter of his own thinking. God shines right on, the mist, never real to the sun, ceases to be real to the lake; the sun is accepted again, the relation is restored, man thinks God; sin, sickness and death disappear from man as man reappears in God.

EFFECT OF SUCH ABSTRACTIONS.

This line of thought insisted upon, repeated, reiterated, becomes real, takes possession of the mind; matter loses its mastery. God is realized as the life of life, and God's man walks in the light as he is in the light. The form, foreign English, the substance, a revelation, make it hard to understand Christian Science. Possibly if the church were more spiritually minded, if the bride of Christ were less at home with the world, if Christians were leading a life that demands supernatural spiritual power to account for it, there would be less place in the world for theosophy and Christian Science.

"The world is too much with us. Late and soon
Buying and selling we lay waste our powers."

Most of the disciples are face to face with unconquered demons at the foot of the mountain; those in the transfiguring glory are content to build tabernacles and abide. The cure for Christian Science is not alone clear thought, but a profounder type of spiritual life.

" 'Tis life whereof the nerves are scant, more life and fuller, that men want." Christ came that we might have life, and have it more abundantly, but the average church life is not an abundant spiritual life.

Not a few worship at the altar of the unknown God, grope after if haply they may find, do not realize that God is "nearer than thinking, closer

than hands and feet"; that he is a "very present help in every time of trouble"; that he is
the God of the body as of the spirit, of the mind
as of the heart. Very few of us realize with Paul
death to the world and life in Christ, and live the
life that is "hid with Christ in God."

There are certain

OBJECTIONS TO THE SCIENCE.

1. It cannot well be a science and a revelation
too; the two words are mutually exclusive. Science deals with truth learned by the use of human
powers. Revelation gives direct insight without
the labor of experiment.

2. Granting that Mrs. Eddy was healed directly
by divine power, her system of thought no more
explains it than thunder explains lightning. Her
experience of direct healing is shared by many,
her explanation throws no light on the fact; it
adds about as much to the truth as the spider's
web adds to the beams in the barn; it gives her
mind a place to swing from, catches flying motes,
and in the dimness seems a part of the beam till
you trust your weight to it.

3. Because we do not know what matter is, it
does not follow that matter is not; because
we do not know what matter is to God, it
does not follow that matter is nothing to him.
We may presume that matter is not to God
what it is to man; it is our limitation; it may be
his expression, and when we have passed through
the caterpillar stage it may prove to be wings to

us in turn. The egg floats and enspheres the germ of life, but under the brooding warmth of the bird the sea of white and yellow becomes building stuff, the shell yields, a winged song comes to the light, the narrow horizon of the shell becomes the wide horizon of the world, giving unlimited play to every power. It may well be that man's spirit is but the germ in a sea of matter, shelled by matter, but under the brooding of the Spirit to build body and wings that shall bear the singing soul aloft into an unlimited future.

4. Sin is a mood of mind, a set of the will, the choice of self; the *act* of sin is but the thunder clap, the *fact* of sin is the lightning stroke that scars and blights. Faith is a mood of mind, a state of heart, a surrender of the will; the soul acts as really when it acts against God as when it acts with God: rebellion is as real as surrender, an impure thought as real as a pure thought; impurity of the heart is as real as impurity of the body, and impurity of the heart is as real as purity of the heart.

5. The assertion of revelation is not proof of a revelation. Christ's full speech fulfilled the broken speech of the prophets; if his teaching had contradicted all that had gone before, it might well have been challenged. This revelation (?) does not explain, it contradicts what has been previously given. Such a method of interpretation applied to law, medicine, business, literature would upset the world. Alice in Wonderland

stepping through mirrors would be a sane guide compared to the leadings of this Wandering Jew of modern literature called "Science and Health."

THE END.

www.ingramcontent.com/pod-product-compliance
Lightning Source LLC
Chambersburg PA
CBHW021939190326
41519CB00009B/1065